I0558337

Lavender Business

Become Your Own Boss Selling Lavender

(Farm Guide to Growing Lavender Plants for Massive Profit)

Timothy Hodgson

Published By **Region Loviusher**

Timothy Hodgson

*Lavender Business: Become Your Own Boss Selling
Lavender (Farm Guide to Growing Lavender Plants
for Massive Profit)*

ISBN   978-1-998038-84-8

No part of this guidebook shall be reproduced in any form without permission in writing from the publisher except in the case of brief quotations embodied in critical articles or reviews.

Legal & Disclaimer

The information contained in this book is not designed to replace or take the place of any form of medicine or professional medical advice. The information in this book has been provided for educational & entertainment purposes only.

The information contained in this book has been compiled from sources deemed reliable, and it is accurate to the best of the Author's knowledge; however, the Author cannot guarantee its accuracy and validity and cannot be held liable for any errors or omissions. Changes are periodically made to this book. You must consult your doctor or get professional medical advice before using any of the suggested remedies, techniques, or information in this book.

Table Of Contents

Chapter 1: A Simple Guide For Growing Lavender For Beginner

Lavender has been substantially grown in gardens for its natural perfume and its use in arts and crafts. Growing lavender has turn out to be famous because of its use in hundreds of aromatherapy merchandise. Lavender has a candy floral heady scent with a woodsy form of herbal aroma to it and is fantastic for supporting to ease anxiety and tension.

It can assist to relieve burns and simplicity headaches as well. Given its large kind of makes use of, a number of people are starting to extend lavender for their very personal private use in growing a mixture of healthy products.

Used in aroma tub salts and bath oils to provide a cushty relaxing bathtub. It is also applied in sleep-inducing merchandise which includes aromatherapy pillows and sleep mists. These merchandise assist you fall asleep thru spraying the Lavender on the sheet. Lavender additionally has many medicinal uses. It can be used as an anti-inflammatory, an antiseptic, an insect repellent, and might carry alleviation to insect bites.

Growing lavender in all fairness smooth and can be finished from quite a good buy everywhere due to the fact it can be grown both out of doors and interior. There are 3 predominant forms of lavender, identified with the beneficial aid of vicinity, which include the French range (Lavandula x intermedia), English variety (Lavandula Angustifolia), and Spanish lavender (Lavandula stoechas). Of the 3, the English range is the maximum typically used whilst developing lavender for aromatherapy, fragrance, and medicinal functions.

Always use sandy soil while developing lavender interior or in bins. Adding lime to the soil also allows to grow healthy and fragrant lavender. When grown interior, you have to placed your subject in an area wherein the plant can get masses of air but it ought to be included from sturdy winds. When watering lavender vegetation, it's miles handiest to dry them shortly in advance than the subsequent watering.

When growing lavender out of doors, sandy rocky soil and sunny locations work tremendous; regardless of the reality that lavender adapts well to pretty a few taken into consideration certainly one of a kind soils. If the usage of mulch, sand, or pea pebbles, make certain to devise the lavender in which you'll get applicable drainage to save you the roots from turning into too moist.

Growing lavender may be an clean and profitable corporation and in all fairness easy even for folks that are not gardening experts. Just preserve the fundamental regulations in mind and also you want to have great achievement developing lavender.

GROWING LAVENDER FOR HIGH PROFITS

Commercial lavender farming is a whole ancient and well-known initiative in lots of regions across the world. It`s clean to begin lavender production and it`s a reasonably worthwhile enterprise for being profitable.

Lavender is a lovely herb and is pretty smooth to develop below the proper cultivating conditions. You may additionally want to make loads of cash from business lavender cultivation. Lavandula is a genus of 47 recognized flowering plant species of the mint own family, Lamiaceae. It is native to the Old World and originates in Cape Verde and the Canary Islands, and from Europe to North Africa and East Africa, the Mediterranean, and from Southwest Asia to India. Many individuals of the genus Lavender are broadly cultivated in temperate climates as ornamental plant life in gardens and landscapes. It is also used as a cooking herb and is commercially used to extract critical oils.

The most generally cultivated species is the blue flower lupine, regularly referred to as lavender, which has a colour named after the shade of this species of flower. However, lavender has been used in traditional medicine and cosmetics for masses of years. And a restricted form of clinical studies resource the healing use of lavender for ache, heat flashes, and postnatal pain. However, developing lavender commercially can be a notable industrial enterprise to make coins. You can sell clean and dried lavender, or make loads of merchandise which encompass aromatic sachets, lotions, teas, soaps, drugs, critical oils and fragrances. Let's take a higher have a observe this industrial business employer.

Lavender is used for plenty unique capabilities is a multipurpose plant. People use lavender in lots of outstanding techniques to foster specific fitness and well-being. Common uses of lavender encompass:

Lavender has long been used by hundreds of people inside the past as a totally wholesome and restoration herb. It is used to useful resource a few clinical conditions and to relieve signs and symptoms and symptoms and signs and symptoms.

Current medicinal professionals have said more than one of the fitness advantages of lavender, but some are however learning to get greater advantages Lavender offers. Nevertheless, I will try and present the precept health blessings Lavender has been supplying for the purpose that days of antique.

NUMBER ONE: It has been showed that Lavender permits human beings affected by insomnia and different sleep issues! All they want to do is stuff their pillows with it and they'll fall asleep.

In the olden days, humans loaded pillows with lavender plants to assist them to get a higher night time's sleep and go to sleep. And currently in 2022, research show that individuals who inhale the fragrance of lavender can enhance sleep incredible.

NUMBER TWO: It has additionally been showed that the overall use of lavender oil may additionally additionally help anyone in treating an undesirable illness known as alopecia areata (This causes the human hair to fall out in spots).

NUMBER THREE: Lavender aromatherapy has been proved to help lessen pressure/tension after a person undergoes the most cancers treatment manner. The calming strength of lavender can heal anybody suffering from broken pores and pores and pores and skin.

NUMBER FOUR: Furthermore it has been proved that the almighty Lavender has a sedative have an effect on, which may be sufficient to alleviate all people affected by migraines and complications.

NUMBER FIVE: Lavender consists of infection-fighting compounds called linalol and linalyl acetate. And for these compounds, lavender merchandise allow you to in treating sizable pores and pores and skin problems collectively with itching, psoriasis, dermatitis, rashes, and eczema. Lavender has extended been used as a time-commemorated burn remedy. Lavender oil can kill micro organism and is a mild way to cope with zits. Some older analyses endorse that there's scientific records to help this use.

THE ADVANTAGES OF LAVENDER FARMING BUSINESS

The business lavender farming industrial organization has numerous advantages/blessings. The first is that it`s very easy, and reachable. Above all, it's miles an simple agency.

It is awesome to apprehend that the commercial cultivation of lavender is a famous business organisation in some global places of the arena. Several individuals who are inquisitive about farming are presently doing this organisation commercially to make a earnings. And the high-quality trouble is that you don`t ought to be troubled too much while you're about to start the Lavender business corporation on account that there can be a large hazard of you making a enormous earnings due to its profitability. Furthermore, the cultivation of lavender might be very clean and it's far beginner's fine! Regardless, it is a appropriate concept to find out about this industrial organisation from specialists...Current farmers who've a first-rate revel in on the subject of taking walks the Lavender enterprise commercially.

Also, the capital necessities are without a doubt low in evaluation to different flowers cultivated for his or her income, so be

9

confident the profits for Lavender are breathtaking.

Also, some other super motive why you have to begin the Lavender organisation is that each the value and get in touch with for for lavender products are excessive inside the marketplace. So you do now not need to be about product advertising and advertising.

To crown all of it, looking after your lavender plant life can be very sincere and clear-cut.

THE SOIL PREPARATION FOR LAVENDER GROWING

You want to perfectly put together the soil in case you need to go into the Lavender commercial business enterprise. The soil steering is distinctly vital for enterprise lavender cultivation. To get commenced, all you want to do is to prepare the land via appearing 2-three deep plowings. For you to acquire this, all you need to do is to put off weeds and bring the soil to the exceptional tilth degree.

Add as lots natural fertilizer or compost even as getting geared up the land. You can add fertilizer from a well-rotten farm or compost to beautify the fertility of the soil.

SITE SELECTION

For Lavender for cultivation, a light, perfectly well-aerated dry calcareous soil with natural-rich slopes is considered first-rate for lavender cultivation. The great soil pH for getting higher important oil degrees from 7 to 8.Three. Also, it's far tremendous you understand lavender flora are very at risk of floods, however the maximum inexpensive detail approximately lavender is that it can expand on bad or eroded soil.

CLIMATIC REQUIREMENTS

Lavender is a sensitive plant that resists drought and frost. The best climatic requirements for developing lavender are cool winters and cool summers. Lavender flowers require precise daytime. The better the altitude, the better the yield you ought to count on. And in case of insufficient lighting fixtures concerns, you will be conscious lessened critical oil content cloth cloth and less yield of plants.

CULTIVATORS/VARIETIES

There are numerous sorts for commercial manufacturing of lavender. Famous and elegant cultivars in commercial organization production embody lavender, English lavender, French lavender, Spanish lavender, spike lavender, wooly lavender, and Lavandin. Most of these types have unique pink buds, however there also are pink lavender, white, and red lavender.

In addition to all I honestly have mentioned above, You can speak over with an current manufacturer in your area to get higher suggestions at the equal time as picking the proper variety on your production.

PROPAGATION OF LAVENDER

Propagation of lavender may be done in hundreds of special methods. But, you have to understand growing Lavender from seeds can be very difficult. Because the seeds typically require 5 weeks of bloodless stratification earlier than you move ahead to plant them. And need spherical 6 months before the seedling sprouts to transplant length.

So, the right method for lavender propagation is taking cuttings from present day plants or looking for lavender seedlings.

Experts endorse taking cuttings from stems (with out a flower buds) after the lavender plant want to have bloomed. Remove the leaves from the bottom 1/2 of of of the plant (lowering), and feature the lavender (reducing) into vermiculite or a properly-draining sterile potting soil.

 Keep the cuttings appropriately watered, and they ought to root in about 3 weeks.

The next trouble you need to do is for you to transplant those rooted lavender cuttings into 2- to four-inch pots. Once they've fashioned healthy roots, then you could move in advance to plant the lavender seedlings in your cute garden.

PLANTING

Lavender vegetation like enough area to increase. So, counting on the Lavender kind you pick, space them 2 -three feet aside inside the row with 3 -6 ft among the rows. And you ought to plant the seedlings in entire solar.

It has been confirmed that the plant density of 20,000 in step with hectare can provide the very first-rate yield of Lavender. Earthing up of soil has to be finished at some point of the seedlings for root popularity quo.

CARING

The lavender plant typically develops effortlessly without a whole lot of help. Although, taking more care will resource the plants in developing properly and generating greater. Some of the common being worried strategies are listed underneath.

FERTILIZING

The lavender crop typically responds thoroughly to fertilizers. The lavender crop normally reacts thoroughly to fertilizers. The recommended K, P, and N, are 40kg, 40kg, and 1000kg consistent with hectare.

Use the whole dose of P and K (20kg as a basal utility, then 20 kg of N. The ultimate 80kg of N need to be applied in 4 splits (2 doses for the length of every twelve months).

WEEDING

Preventing weeds can be essential in the lavender farming business organisation.

Because weeds consume vitamins from the soil and the lavender plant life will go through.

Regular weeding and hoeing are needed to keep the planting floor weed-unfastened.

1-2 times hoeing is demanded after thirty to fifty days of sowing to govern the weeds. Mulching is also a effective way to reduce soil temperature along aspect weed control. Flower buds of lavender plant life must be pruned off within the direction of the primary 2 years period to allow the flora to amplify a robust framework.

WATERING/IRRIGATION

For business lavender agribusiness, presenting enough watering or irrigation may be very essential. In commercial organization manufacturing, irrigation want to be furnished for 2 years till the crop has been hooked up.

Irrigation need to be introduced at critical stages of plant increase in the case of lighter soils and coffee rainfall areas.

You need to make sure you observe the proper irrigation at some degree within the flower initiation. Try to avoid overhead irrigation (e.G. Sprinkler irrigation) as it may improve disease troubles. Try to undertake drip irrigation alternatively, because of the truth this tool can preserve water and manipulate weed growth.

OVERWINTERING

Having evolved along the Mediterranean Sea, lavender is not familiar with snow and bloodless weather. If you've got got difficult weather conditions, protect your lavender flora with heavy mulch, wind blocks, and cloth row covers.

MULCHING

Specifically, natural mulches aren't right for lavender, due to the plant`s vulnerability to fungus and molds. When you mulch your Lavender with white stones or white sand, then you will be boosting the plant's chances of manufacturing oil and flowers at the same time as keeping down weeds and reducing fungus infections. Black landscape cloth is a ultra-modern weed barrier for cultivating lavender plants.

PESTS & DISEASES

Very few pests upward push up on lavender and no severe ailments are suggested in the lavender farming industrial enterprise. Although, you can visit a professional on your place in case you be conscious any problem.

HARVESTING

Typically, the lavender vegetation begin flowering earlier in low altitude and warmer areas and flowering starts offevolved late in immoderate slopes. As a part of harvesting, you need to lessen the flora whilst the stem duration is round 10cm.

Lavender is superb harvested even as spherical 1/2 of of the flower buds have opened. You can use sharp scissors for harvesting lavender through the stem (truely below the number one set of leaves) within the morning at the same time as the oils are the maximum targeted.

PRESERVING, STORING, AND DRYING LAVENDER

There are endless methods you may choose out to package deal or keep your lavender if you`re making plans to put it on the market it on the farmers' market or via wholesale. Perhaps the precise technique is dried lavender.

Chapter 2: 9 Things You Need To Know To Grow Lavender

Lavender, famous for its sweet-smelling perfume and crucial oil, is a should among herb enthusiasts, or maybe amongst rookie herb growers. After all, can everyone oppose the kind of sensitive floral fragrance? But one pressing question is: Is it easy to develop? Most gardeners will let you know that lavender is a drought-tolerant herb that prospers high-quality in warm and humid situations. But this is most effective right as soon as the plant has set up its root tool already.

In the early developing duration of lavender, care want to just accept to it till it reaches the length even because it will become the hardy plant that it is acknowledged for.

So in case you are yearning and craving to grace your herb garden with the ones cute herbs, right here are some useful recommendations that you need to undergo in thoughts to make sure which you obtain fulfillment with lavender herb gardening:

NUMBER ONE

Unless you need it, do now not broaden lavender from seeds, and that could be a very sluggish and tough manner. First off, lavender best grows sufficiently from glowing seeds, and coming via easy lavender seeds is tough. If you`re fortunate to ultimately find out precise seeds, it's going to take severa months to develop roots and bud the number one few leaves. Many conservatories offer lavender flowers, or in case you understand each person with a grown plant, you could even ask for lavender cuttings, that would take roots in quick after it's far planted.

NUMBER TWO

Lavender is a solar-lover herb and prefers barely arid developing conditions. You want to commercial enterprise company to generally plant it in the sunniest spot in your lawn.

NUMBER THREE

Its roots are sensitive to moisture and do no longer like soggy clay soil, which holds too much water. Mix sharp sand and correct compost into the soil to decorate soil shape and drainage. The compost may additionally even offer vitamins to useful resource the brand new lavender plant's root improvement.

NUMBER FOUR

You can growth lavender in a container, however do now not choose a big pot because it adapts very well to tight regions irrespective of its exceptional root device. Make superb the container has appropriate enough drainage and use sandy soil blended with compost.

NUMBER FIVE

When planting on the ground, maintain the plants at the least 2 toes aside to allow sufficient air move between the plant life. Humidity is some different stressful hassle with lavender and may be averted if the plants aren't too close.

NUMBER SIX

You do not want to water the newly planted lavender besides you are planting it in dry soil. Normal and slightly wet soil will deliver enough moisture for days. But new vegetation may additionally call for extra ordinary watering than the matured flowers, despite the fact that you may however without problems overwater them. To prevent this, water your lavender exceptional while 2 to a few inches of the topsoil is dry. Don't water from above. If you are developing lavender in boxes, soak the containers in a bucket full of water until the topsoil turns into wet. Then drain right away. As the plant matures, it needs a first-rate deal a great deal less watering.

Chapter 3: 7 Methods To Make Money Cultivating Lavender

Lavender may be one of the maximum useful and powerful coins vegetation for small farmers or farmers who're beginners. Even a patio or small outdoor lavender garden can produce an first rate amount of profits if completed well. Best of all, in contrast to different seasonal produce, which include plants, which might be useless if no longer advertised at harvest time, you may dry the specific lavender and make it into even extra worthwhile products. Here are 7 of the maximum suitable techniques to reveal lavender into cash.

I. Dried Bouquets

Unsold lavender bunches may be dried thru putting the other manner up and bought to florists and artisans, who use the bunches for dried flower arrangements. Also, the stunning flower buds of the lavender plant may be extracted from the bunches and bought (or used to make sachets) and distinct rate-introduced through-merchandise.

II. Fresh lavender bouquets.

For novices or small growers, that is a promising manner to sell lavender. Most planters promote direct to the retail loads, both from their patio lawn or on the close by farmer's market. At the provincial Saturday market near my domestic, lavender bunches promote for $five-$6 each. A 20' x 20' developing region can produce round 2 hundred-3 hundred bunches every 12 months, well worth $2,000+. More high-quality plots are even extra useful. 1/4 acre can beget about 3,000 bunches, properly properly well worth $20000+.

III. Dream pillows

Lavender is thought for its soothing impact, so putting it in a pillow makes feel to assist foster restful sleep, and is one of the maximum beneficial charge-brought lavender by using way of-products. Medical analyses have even decided that lavender can help calm youngsters with ADHD. One bold lavender grower has designed a line of animal-themed dream pillows for kids, grossing over one million dollars every 12 months from her pillows.

IV. Sachets

You can use Lavender sachets everywhere the air needs deodorizing or freshening, together with drawers, wardrobes, even in smelly footwear! Most sachet offers come from repeat customers, who treasure the scent of lavender. Sachets also are marketed to local sellers. The Saturday market is a notable spot to promote sachets, especially if they may be made the use of decorating material scraps.

V. Pet Products

There are severa promising home dog merchandise that can be made from lavender, however a lavender flea repellant is constantly a awesome-provider. Many flea repellants product in the market use sturdy chemical materials which can have toxic factor results. Lavender is an all-natural flea control that no longer outstanding appalls fleas however also makes pets fragrance better!

VI. Live vegetation.

Most lavender farmers discover promoting stay plant life a totally profitable addition to their lavender corporation, because of the truth the returns are large. To make sure the lavender plants are particular duplicates of the determine vegetation, planters, take root cuttings from them, in place of growing from seed. The handiest rate is for potting soil and pots; with named types in a 4-inch or 6-inch pot brings $5 and further. Farmers can even wholesale potted live lavender plant life to provincial garden centers and conservatories in their place.

VII. Lavender cleaning soap.

Many clients don't forget lavender cleaning soap is important in the washroom. With such a whole lot of molds to be had to soap makers, lavender bars can be made in an almost endless shape of shapes and sizes. Of course, cleansing cleaning cleaning soap is a repeat product and a very well-known present object. New cleansing cleaning soap makers can get triggered effortlessly via way of using a "soften and pour" cleaning soap base.

As you have seen, lavender planters have numerous paperwork to make cash from this brilliant plant. By the usage of fee-brought products, the profits from a patch of lavender can maintain 12 months-spherical as opposed to being confined to sincerely harvest time.

Chapter 4: Why The Lavender Industry Is Booming Around The World

When you concentrate the word "Lavender," you may possibly hyperlink it with numerous subjects. It's a huge accent designed alongside bouquets and flower arrangements. For a few human beings, it comes with reflective recollections just like the calming fragrance in their conventional home. You may also furthermore recollect the spa and massage, or you'll probable think about its scented competencies. Lavender cultivation spherical the area will boom in popularity because of its many blessings.

Lavender has set up to be an outstanding useful resource for its fragrant, healthful, enjoyable, and distinct well being benefits. This financial ruin will deal with all approximately Lavender agribusiness and why this organization maintains to increase these days.

Are you aware numerous thru manner of-products we use in recent times hire lavender as a primary element? You need to understand a Lavender Pillow will provide you an increased degree of ease while you want to take a brief relaxation or an extended night time's sleep. Almost anybody round the arena adores the heady scent of the almighty lavender, so in recent times; markets round the arena are full of smells and sterile through-merchandise which have it.

LAVENDER'S RISING CONSUMER DEMAND

France is the number one company of Lavender and this is why over 30% of the complete worldwide manufacturing of Lavender comes from France. During summer season, you should anticipate to look several fields of lavender within the southern region of France. Unfortunately, there has been a decrease inside the deliver due to climatic and environmental problems. But regardless of all this, the call for throughout the place from customers maintains on developing. And due to that, the experts in the organization are continuously searching out techniques to generally meet the immoderate demand for Lavender around the world.

France's Lavender farmers persist to inspire neighborhood farmers to expand Lavenders and supply unique enough belongings. They maintain to include techniques to boom supply, it is a sufficient justification for why they live a leader in the lavender region. In addition to France, many nearby farmers round the world have began to broaden lavender for cash due to its want. The plant is a favorite among consumers, so so long as growers and carriers efficaciously control its manufacturing, the company will keep to increase, making it a required useful useful resource within the marketplace these days.

Chapter 5: Various Uses Of Lavender

I. Aromatherapy

 Lavender flora are purported to have a non violent heady scent that stimulates relaxation. Aromatherapists often use lavender for its medicinal results. And for this encouraging reason, nearly every person is reminded of rub down spas and parlors after they have a sniff of Lavender. In addition to relieving anxiousness and pressure, the almighty lavender also can furthermore lessen minor pain.

II. Sleep Support

When it comes to sleeplessness or first-rate sleep situations, Lavender is believed to be very useful. To get a exquisite night time time's rest, human beings of old may also insert stalks of dried lavender into their sleeping pillows. This method isn't previous whilst you preserve in mind that there are notwithstanding the reality that severa enjoyable pillows that encompass lavender.

But there are also different way to ensure a tremendous night time's sleep, just like the use of cushty and comfortable pillows which includes the Hemp Pillow.

In addition to allowing you to sleep and lighten up, you can use Lavender oil to loosen up the complete body, this is why it is a set up preference amongst rub down therapists. There are such pretty some via the use of-merchandise that use Lavender as their primary element. You can use the Lavender oil as a natural sleep useful useful resource, and you could moreover use it in a diffuser in advance than bedtime.

III. Helps relieve Pain in Dementia and Cancer Patients

Aromatherapy (the use of Lavender oil) is positioned to be precious in some maximum cancers patients. In handling ache and side impacts of maximum cancers remedy. Those who are affected by dementia may additionally additionally locate aromatherapy treatment beneficial.

IV. Hair Support

Alopecia Aerate, an inflammatory sickness that impacts the hair can be dealt with with Lavender oil topically. Other not unusual important oils like Tea Tree, Rosemary, and Peppermint can be implemented in treating the same hair hassle. If you're present process some pending problems together with your hair, you need to try and observe herbal oils topically, and you may see some powerful results thereafter.

V. Skin Treatment

Lavender oils have been used to mitigate numerous pores and skin issues collectively with pimples, eczema and sunburn. With the extensive form of pores and skin care products in the marketplace nowadays, you may be capable of locate exquisite best ones which have lavender extracts.

LAVENDER OIL - AN INCREDIBLE VARIETY OF USES

Lavender oil has long been cherished no longer best for its reinvigorated, candy perfume but moreover for its restorative outcomes. Its file is going decrease back at least as a protracted manner due to the fact the Roman Empire in which lavender changed into used to freshen the wash water and the oil have become used to deal with war injuries. In modern times lavender oil has numerous uses from aromatherapy to heady scent for tub thru way of-merchandise. Learn approximately this tremendous herbal substance from plant to spinoff.

What is it? - Lavender oil is the crucial oil of a lavender plant, emanating from the vegetation and stalks (peduncles). The plant material is subjected to a steam distillation technique that produces lavender oil and hydrosol.

WHERE DOES IT COME FROM?

Lavender thrives nicely in a huge type of climates and may be positioned in masses of areas of the earth. It matures wild in severa Mediterranean global places, and there are numerous lavender farms in this vicinity. France is famous for its lavender farmsteads and festivals. Other massive exhibition locations encompass the usa, Australia, and New Zealand.

How is it used?

The super form of blessings for this extraordinary oil is terrific but may be labeled into large lessons: Scents and Therapeutic.

Therapeutic Uses- Lavender is probably the maximum everyday essential oil utilized in aromatherapy, it's an possibility fitness remedy that makes use of crucial oils. Breathing the aroma of lavender has been validated to have a relaxing effect on many humans and is frequently used for sleeplessness. The aroma is again and again circulated the usage of a burner or diffuser, however you may get pride from the usefulness with the useful resource of in reality setting some drops of lavender oil in a cotton ball to take a seat down in your nightstand or on your bathwater.

Lavender oil has ache-killing outcomes and supplies brief remedy from hurting joints. Rubbing the oil proper now into can help relieve moderate arthritis or youth "growing pains." The antibiotic consequences of lavender make the oil a great natural opportunity for thwarting contamination in burns or minor cuts. Simply use the oil right now to the wound and feature a laugh with each the ache-killing and germ-killing consequences.

Lavender Scent

Lavender has an exceptionally candy scent and the oil has been used for hundreds of years if now not masses of years as a nice aroma. The fragrance of lavender may be determined in severa by using the use of-products, from air fresheners to soaps to lotions. Lavender is a desired aroma in spas and is a first-rate piece of many facial creams and massage oils. Lavender salt scrubs and sugar scrubs are terrific approaches to exfoliate the pores and skin whilst relishing the soothing fragrance.

These advantages of lavender oil are but a number of the severa strategies that you'll be capable of take gain of the splendid aroma and fitness blessings. It is one of the amazing gadgets of nature.

Chapter 6: How To Harvest Fresh Lavender: The Steps To Harvesting, Pruning & Drying Lavender Flowers

Are you cultivating lavender to your outdoor or lawn? In this Chapter, I might be education you the way to reap clean lavender flowers, at the facet of numerous strategies to dry and use them! Of all the fragrances in our lawn, the scent of glowing lavender blossoms need to be considered one of my all-time dearests. Yet lavender is a lot more than only a lovely, fragrant flower! Relieving to your mind and pores and skin, lavender is an fit to be eaten, adaptable, instead-healing plant. It is also very appealing to pollinators, that is usually a heat function in a sustainable garden.

English, Spanish, French...oh my God! No undergo in mind what form of lavender you plant, all of those suggestions on a manner to reap and dry glowing lavender may be very useful to you. Harvesting lavender is quite clean. Additionally, the more you harvest, the more blossoms will come! In this bankruptcy, we'll move over the excellent period to achieve lavender, precisely wherein to trim it, similarly to the manner to deliver the plant a higher prune. Then I'll display to you 3-strategies to dry glowing lavender buds, and percent an abundance of opinions on what to do with them! How do lavender salve, sachets, and cocktails sound?

SIMPLE TIPS FOR GROWING LAVENDER

Cultivation of lavender is notably smooth, given the right conditions and climate. Lavender has the capability to increase as a perennial in USDA Zones five-10. Lavender flourishes in a warmness, sunny, dry Mediterranean weather and does now not artwork nicely in damp or humid conditions.

One manner to balance everyday rain and humidity is to develop lavender in a field. Use nicely-tired potting soil mixed with sand or cactus soil. Above all else, pass overwatering it. Overwatering and Excessive moisture is the maximum ordinary reason of lack of life in lavender. Soggy roots will speedy bring about fungal infection and extinction. When planting lavender proper now in the floor, select a sunny area with best drainage and sandy soil. Lavender doesn`t want a whole lot of compost, fertilizer, or noticeably wealthy soil. It feels most comfortable on rocky ground.

Lavender is feasible but sluggish and grumpy to start with seeds. Follow the ones instructions to plant lavender seeds interior, however provide them as much as a month or two to broaden. The quickest and best manner to growth lavender is initially an established nursery or seedling. In fact, we have were given finished it always! Be affected character, even more youthful flowers can develop slowly inside the beginning. In the second one or 1/three 12 months, the roots settle, develop and bloom greater actively. Sleep in the first 12 months, circulate slowly in the 2d 12 months and bounce in the zero.33 three hundred and sixty five days.

LAVENDER VARIETIES

There are large types of lavender to pick out from. Most of them fall into one of the more favored education of Spanish, English, or French lavender. Some live pretty compact, on the identical time as others get massive! Dozens of hybrids exist too. All sorts offer a super scent and are technically secure to eat, in spite of the truth that lavandin sorts (L. X intermedia) and English kinds are the most treasured and match for human consumption for culinary packages.

You need to are seeking out types which may be well-appropriate in your weather. For instance, actual English Lavenders (Lavandula angustifolia) are the maximum cold-hardy, together with 'Munstead' and 'Hidcote'kinds of English lavender. Those are the few rated down to area 5. Spanish and French lavenders do better in warmer climates and higher zones. 'Phenomenal' is a actual English lavender that is regarded to bear the warm temperature and humidity of the southeast. Lavandula stoechas and L. Dentata are also amongst some of the extra humidity-tolerant kinds.

If you aren't powerful what your USDA growing sector is, use the clean zip code research device to find out. I continuously suggest plant shopping at domestically-owned nurseries – they'll probably convey lavender types pleasant conformed to your place! We expand a large preference of English, French, and Spanish lavender sorts in our touchy garden.

A high-quality Spanish lavender plant is featured with loads of sparkling crimson blooming plant life. The red contrasts the plant's mild mint-coloured foliage.

THE STEP BY STEP PROCESS TO HARVESTING LAVENDER FLOWERS

43

If you have perfected your Lavender growing technique, you should understand one of the maximum on hand (and profitable) additives is harvesting. You ought to understand that the greater you prune (your Lavender), the greater it'll maintain blooming! Regular pruning and harvesting of your lavender will enhance branching, as a manner to in the end cause a bushy plant.

In my 10 years of growing Lavender, some of the maximum not unusual inquiries I accommodate from novices approximately developing lavender are "whilst is the maximum suitable time to benefit lavender?" and "How they're able to with a chunk of good fortune gather or prune lavender faultlessly?" – So allow's talk this on this Chapter!

WHEN IS THE MOST SUITABLE TIME TO HARVEST LAVENDER

As a professional in cultivating Lavenders, I would possibly say the maximum appropriate and suitable time you may harvest lavender is: early, on numerous ranges! All I can say is early bloom, early morning, and early spring is best for harvesting Lavenders.

When you pass in advance to reap your lavender flora inside the early spring, then you definately without a doubt may be giving the plant sufficient time to beget another flush of blooms you may enjoy over again(late summer time to fall). This is specially relevant if you have a quick length of summer increase. Also, you have to apprehend in frost-loose climate wherein a few sorts of lavender can also moreover moreover blossom all twelve months round, you can typically harvest small corporations time and again throughout the year.

We all recognise the fragrances from Lavenders are continuously soothing, so when you have yearnings for attractive perfume and crucial oil content material, the maximum suitable time to gain character lavender flowers is early in their bloom cycle. This indicates that the soft more youthful shoots are sturdy and without a doubt beginning to flower. Fully opened lavender vegetation display most of the color and are very appealing for bouquets. As lavender flora mature, their healing oil and aroma content material material decreases. As the lavender vegetation mature, their healing oil and aroma content material decreases. Also, mature brown flower buds collapse more resultseasily and fall off the stem. This isn't maximum fulfilling for bouquets and can lessen to rubble the drying cycle. Regardless, I don't allow that save you me from occasionally the usage of older blossoms too! Late is higher than by no means, as pulling spent blooms is appropriate for the plant even though.

Ultimately, herbalists commonly collect recuperation vegetation early inside the morning, at the identical time because the flowers are but active from the chilly night time time time air. As the day progresses, a number of the fragrant vital oils and terpenes will start to disperse within the warm sun. I try to keep this workout, especially once I want to make medicinal oil or cream from the lavender buds.

WHERE TO HARVEST LAVENDER FLOWERS

To harvest separate lavender plants, you need to first pinpoint the bud which you choice to attain. Then, you could need to comply with the stem down from the flower bud till you get to an area wherein new buds,   thing leaves, or branches have started out to form. Using little pruning scissors or snips reduce the stem there – sincerely above the aspect branches or leaves. Once the middle flower and stem are removed, the plant diverts its power. Now, the ones element shoots will swiftly make bigger and supply sparkling plants of their very very own! Once the center flower and stem are removed, the plant diverts its strength. Now, the ones factor shoots will fast growth and supply easy plant life of their very private!

For more prolonged bouquet or stems that also possesses a few green foliage, actually examine the precept stem down a hint farther, snipping above an same branching junction but greater in-depth into the plant. You can also discover the need to do that with smaller, narrower lavender plant life that have a shorter distance among leaf nodes or buds. Or, to prune a piece over a tall, mounted plant.

After harvesting your lavender, you'll be left with a pleasant little bunch of lavender – this is fine to dry and dangle, or to reveal as a cute bouquet.

Chapter 7: The Step-By-Step Process For Pruning Lavender – And The Most Suitable Time Of Year To Do It

Do you realize it is important you recognize the manner to prune the Lavender? If you preference these sacred aromatic vegetation to live in suitable form for destiny years, you need to discover ways to prune it perfectly.

Lavender is a completely essential plant if you want to add perfume to your lawn and has prolonged been valued for hundreds of years for its medicinal possessions. It's additionally an appeal for butterflies, bees, and distinct loved pollinators, building a sanctuary for flowers and fauna.

Nevertheless, if left on its very very very own, lavender can turn out to be ragged and woody. If neglected for too lengthy the flowers will labor to get well and also you'll want to replace them as-soon-as-feasible with new ones.

'Don't be scared to prune lavender — the Lavender plants can emerge as woody and leggy very rapid, and practical pruning will expand their lives,' says expert!

The fascinating news is that it's easy to recognise the manner to broaden lavender from cuttings and seed, so in case you are struggling to maintain gift plant life looking great, then you can domesticate a everyday deliver of new ones.

As you understand through manner of now that there may be more than one shape of lavender — English, Spanish, and French Lavender sitting at the top. English lavenders (together with Munstead and Hidcote), are the hardiest and the most famous. Other European lavender types are a great deal much less hardy, and so you should take greater care at the same time as pruning. Nevertheless, if you hold to three golden instructions, you could use them for all of your lavender plant life.

HOW YOU SHOULD PRUNE YOUR LAVENDER IN ITS FIRST YEAR

Lavender requires mild pruning within the first three hundred and sixty 5 days, but it desires to make a promising start to save you the flora from turning into prolonged-legged inside the future.

In this early stage, pruning is the advertising of new boom and the formation of first-class mounds. When lavender is grown from seeds and cuttings, it's far well worth learning new boom thoughts simply so the plant can broaden greater wooden.

You must cope with pruning new lavender at some stage in the summer season after the plant has flowered. Using a hygienic, sharp pair of secateurs reduce every stem lower back thru manner of as a notable deal as a 3rd, to extract the flora and some of the inexperienced stem boom.

Try no longer to lessen the plant once more 'tough' by using undertaking the woody bottom of the stem – it's far critical to move away an abundance of green on the stems while the lavender flowers are younger. Try to create an high-quality domed shape through way of leaving the longest stems in the middle and slowly shortening them as you pass to the outer edges of the plant.

After pruning your lavender, you can get a 2d flush of flowers. Trim the ones, the same way as quickly as completed – however do it properly earlier than the cold fall climate gadgets in.

HOW TO PRUNE ADULT LAVENDER PLANTS

Lavender plants will set speedy, so from their 2d twelve months you can want to have a look at a sincere – but careful – pruning exercising to keep their shape.

When choosing a way to prune lavender, there are numerous colleges of concept approximately whilst and the way regularly to do it. But it's most low priced to attack the plant life in tiers: Prune after it vegetation, then you could prune it inside the spring!

Don't prune lavender too tough after summer time ends, or it could conflict to address the onset of bloodless climate.

If you crop the entire plant returned to antique wooden it is able to suggest a large hassle. Lavender is ever-green, which shows that it holds and dreams its leaves all iciness. If you chop into the antique wood, which does not have any leaves, and new leaves do not grow, then it'll not stay on.'

If you fail to prune your lavender during the summer time, then it's fine to attend until the following spring, specially for a whole lot much less hardy Italian, French, and Spanish lavenders.

When trimming your lavender in the summer time, you can like to reap the vegetation for his or her culinary and medicinal uses. If so, try this whilst about 1/2 of of the buds are in bloom. Otherwise, you need to pause till the flora have wilted before pruning.

To trim your lavender within the summer time, draw close handfuls of the stems and, using tidied, sharp secateurs, snip them off, extracting as an lousy lot as a third of the plant's increase.

Try to keep a properly-rounded shape to the plant, but do not reduce too close to to the woody backside of the stems, or the plant also can stumble to overwinter.

To dry out the Lavender flowers, gather the stems into bunches, bind them together, and adhere them the opposite manner up.

HOW TO PRUNE LAVENDER PERFECTLY IN SPRING

Spring is the excellent time to prune the lavender to scale back the increase of woody stems and inspire sparkling new growth. You need to do this early within the season to provide the lavender plant enough time to reestablish itself.. You ought to try this earlier within the season, to offer the lavender plant abundance of time to re-set up itself.

Take a stem and examine it — you'll see it has a woody backside set beneath the leafy area. How heaps wooden there relies upon at the plant's years, and the manner well it is been pruned.

Use easy, sharp pruning shears to lessen the stem into the luxurious region of the stem, about 2-3 inches above the wooden base. Do now not lessen into timber!

You can do handfuls of stems at a time, and for hedges, you'll likely find it extra cushty to use shears.

Try to make a pleasant round form on your lavender plant with the resource of trimming the outer stems a bit briefer than the internal stems.

Where there are frost-damaged, dead, or diseased stems, those ought to be absolutely cleared.

HOW TO PRUNE YOUR WOODY LAVENDER

When lavender is a few years vintage, it could develop prolonged, 'woody' stems that look unattractive. Nevertheless, it can be possible to rejuvenate the lavenders.

''The popular recommendation is to update plant life after they end up leggy, usually after 3-5 years. But I skip having to try this through slicing right decrease again into the wooden. I truly have no longer mislaid a lavender plant but inside the 10 years they were growing in my lawn.

Though usually dodged, reducing lavender into the antique wooden can be an affordable way to refurbish them. The ploy is to make certain you could nevertheless see signs of life inside the shape of growth nodes absolutely below the reducing element. If you chop beyond this, the stems are a long manner-fetched from recovery, so check the stems carefully.

Bear in mind you're bearing a risk, so earlier than you try to difficult prune woody lavender, take some semi-ripe cuttings, so if your plant fails, you can broaden a brand new one.

THE PERFECT TIME TO PRUNE LAVENDER

The most suitable time to prune lavender is within the spring or late summer time, but maximum specialists recommend pruning intervals a year – a trim submit-flowering inside the summer time and a 2d, greater tough pruning within the spring.

Another famend professional even proposes the third trim inside the fall, to allow it 'hold an first rate pebble form'.

HOW DO YOU PERFECTLY CUT BACK LAVENDER FOR WINTER?

Cut once more lavender in advance than wintry weather to form a tidy mound so you can provide shape to the garden over the freezing months. Lavender is an evergreen shrub, so it keeps greenery yr-spherical. It's most reasonable to do your first prune earlier than the autumn, however hardier kinds can react well to a fall pruning earlier than the wintry weather.

Leaving dwindled blossoms at the plant also can deliver meals to seed-consuming birds, so it's not essential (generally) to eliminate the flora immediately after blooming.

HOW DO YOU CUT LAVENDER SO IT CAN GROW BACK?

For you to cut lavender just so it is able to broaden lower again, it's miles critical to keep away from reducing into the "lifeless'," woody increase. If you harvest lavender without a doubt as it's miles flowering, you may get a 2nd flush of plants.

Chapter 8: How To Dry Lavender Flowers

Now which you have gathered smooth lavender, permit's speak about three particular techniques you could use to dry it. Which method you select is simply as a bargain as you. Let's have a take a look at some of the strengths and weaknesses of every.

I. Hanging lavender and letting it dry

The most effective way to dry glowing lavender is to dangle dry it.

Collect a handful of huge bouquets, tie the stems with a rubber band or string, and keep them the other way up to dry passively.

If to procure an entire lot of lavender at one time, it won't dry speedy. It makes the maximum sense to make numerous small bouquets and maintain near them in choice to one big one. Large, thick chunks of lavender have masses a whole lot much less airflow, are slower to dry, and are greater liable to mould.

For this cause, do now not over tighten the ribbon. The tufts want to be tight enough to prevent them from falling aside, but not too tight in the direction of the stems. Hang a bouquet of lavender in a warmth and dry place with sufficient air go along with the flow. The open domestic windows and close by enthusiasts are superb. Drying the lavender in a dark location (out of natural mild) will decorate shade retention.

The time required for complete drying can range from weeks to a month or extra, counting on the weather. To take a look at if the lavender is dry, try breaking one of the stems. When absolutely dried, it's far more likely to fold in 1/2 than to bend. You also can erect the bouquet and dry it (for example, in a water-free, ethereal vase).

This approach of drying lavender works nicely in warmness and dry climates or in first-rate indoor situations in which lavender is dried. Living near the coast, we undergo a piece of mist and slight humidity, and our ornamental lavender dries as an opportunity properly this way. We have bouquets of dried lavender on show in every room!

II) Dry the lavender in a dryer

Another way to dry lavender is to apply a dehydrator. This technique is likewise very on hand and hundreds faster than passive drying at room temperature. I decided to use a dehydrator to dry the lavender for use within the production of lavender-injected ointments and oils. Not handiest does it boost up the tool, but it additionally lets in to dry the lavender one hundred%!

Insufficient drying of wet herbs may want to make medicinal oils more vulnerable to mold and rotting. In cool, foggy summers, this is precisely what works well to ensure fulfillment. Still, it makes the most experience to keep away from overheating lavender to attain the very outstanding viable crucial oil content material material cloth and recuperation effect. Therefore, set the dehydrator to the minimal temperature putting (heaps much less than one hundred-one zero 5 °F).

To dry the lavender, do the subsequent:

Step 1: First you want to acquire lavender. I want to preserve the plants clean via pulling out the complete long flower stem at harvest, even though I'm surely drying the buds myself. Collect lavender in a package deal deal deal of buds and decrease off extra stems (throw right right into a pile of compost).

Step 2: Next, the lavender buds need to be positioned in a single layer at the dryer tray. If the lavender is mainly brittle, or if the dryer has big openings, it is useful to use a tray liner (if any). Alternatively, if the flower buds have fallen, you can line up the parchment paper at the tray of the dehydrator. Dry lavender at low temperature. Our Excalibur dehydrator has a placing (90 five-a hundred and 5°F) created to preserve healthy plant enzymes. Lavender takes about 24-48 hours to dry sincerely, counting on the size of the bud, the shape of lavender, and the gadget used. To test if they'll be actually dry, strive breaking the large sprouts aside. It feels dry and brittle and need to harm in place of bending the treasured stem.

When the buds are dry, switch them to an hermetic container for storage or one-of-a-kind final destination/use.

III) Drying in a basket or sieve

The 0.33 and very last technique of drying lavender might be very much like the primary technique, except for bundling and setting. Some significant herbalists dry sparkling herbs and plants in a colander or a well-ventilated basket.

This way, you can dry the complete lavender stem or actually the buds. Home-made herb drying frames can be produced from unmarried or more than one "frames" of flat frame video display units. Alternatively, you could use a technical multi-layer putting herb drying rack.

In addition, as with the primary approach, passive drying of lavender with a show or basket requires respectable, dry situations and time. Spread the lavender flora in a single layer to enhance the airflow between them. The wicker basket is complete of freshly harvested plant life. They are several layers deep and need to be extra lightly dispensed for storage.

HOW TO KEEP DRIED LAVENDER SAFE

After the lavender has dried, you could leave the bloom buds on the lengthy stalks. It's first-class for showing dry bouquets and dried flower displays. For maximum freshness, taste, and fragrance, the flower buds are clipped or peeled from the stems and preserved in a closed glass box. Keep the box somewhere cool, dark, and dry.

How to Use Lavender Powder

Aside from the bouquet, dry lavender can be used in severa methods. Lavender has a tremendous form of makes use of and advantages. It has been tested to help alleviate tension, anxiety, disappointment, and insomnia even as applied in aromatherapy. Pests also are advised to be repelled with the useful resource of the usage of the aroma. In addition to its amazing scent, lavender has many particular health blessings. With its herbal anti inflammatory, antioxidant, antibacterial, antifungal, and ache-relieving houses, dried lavender is good for use in domestic made tablets and private care merchandise.

As an in shape for human consumption flower, lavender additionally has many culinary makes use of!

Lavender offers lots of culinary applications as an healthy for human intake flower! Remember that the English and lavandin hybrid kinds are the maximum well-known for consuming. Spanish and French lavender is not aromatic or sweet due to the immoderate camphor degree. Camphor, as an opportunity, is a brilliant terpene for decreasing swelling, itching, and pain.

Here are some tips for a way to use dried lavender.

•Make a relaxing lavender potpourri packet with a small mesh bag or cheesecloth it's miles fantastic for a cloth cabinet drawer, relaxation room, vehicle, or bedside table.

•Use to make lavender-infused oils, tinctures, ointments, soaps, and body washes, amongst various things.

•Make clean lavender syrup. Making a slight lavender syrup to function to homemade kombucha is in reality one in all our favorite activities (I bypass light on the sugar, wealthy on the lavender). In truth, virtually considered one in all our favorite kombucha flavor combinations for the second fermentation is lavender-lemon. Cocktails with lavender syrup are notable!

•Lavender sugar may be made and implemented in cuisine along with sugar cookies, desserts, and cupcakes.

•Lavender works well in every sweet and savory marinades.

The maximum everyday software program software is for meat, in spite of the reality that it could additionally be used for roasted potatoes and awesome vegetables.

•To repel insects, lower odors, and soothe chickens, sprinkle dry lavender over coops and hives. We use antique vain lavender shoots and stems as herbal mulch and bug repellent for potted flowers on a every day basis.

Chapter 9: Best Species To Grow

With forty seven one among a kind species, the genus Lavandula isn't genuinely the maximum severa enterprise organisation of flowers available, but it actually has lots of exceptional options to choose from.

There are a few significantly grown by using the use of way of organization growers across the arena, which can be people who I receives into on this financial ruin.

When considering which kind of lavender is proper on your lawn, you'll first want to pay near hobby to three crucial info. Although lavender is an clean plant to enlarge, you want to make certain you choose the right variety to your weather, have properly-worn-out soil, and apprehend the manner to space and trim the vegetation.

Try to determine what you're going to do with the lavender in advance than you plant it. Good corporation and making plans are what gadgets agencies aside. Make advantageous to recall every species cautiously and pick out out the right one, as a few retail clients, just like the ones at farmer's markets, will purchase it for culinary use, a few for sachets, and others for the vital oil.

So, earlier than growing your flora, make sure to do a little marketplace research for your place. What are your opposition selling? What are clients seeking out? What are you able to provide that may be unique from that?

This will offer you with a reasonably accurate idea of the species you need to be that specialize in.

To help you recognize the one of a kind types you'll come across, right here's a top stage view of the 4 crucial kinds of lavender to be had:

•English Lavender

•Lavandin Hybrids (Lavandula x intermedia)

- French Lavender

- Spanish Lavender

- English Lavender (Lavandula angustifolia)

English lavender is also referred to as not unusual lavender, as it's miles the most typically grown species in the company. It is notion for its delicious aroma and pink plants. They are small, tight flower clusters that bloom in the early part of the season, set toward blue-inexperienced leaves. These hardy lavenders perform properly for northern gardeners, overwintering to vicinity 5. Those gardening in a good deal less warmth zones will need to depend on a hotter microclimate internal their lawn beds to make certain the plant life' survival.

These lavenders are commonly the first preference for culinary gardeners.

Despite its call, English lavender is neighborhood to the Mediterranean. It gets its name from the reality that it grows in particular properly in England's cooler climate.

This type of lavender is usually 2 to 3 toes tall (0.6 to at the least one meter), with slim, gray, and green leaves on square stems. The bluish-purple vegetation seem in overdue spring to early summer season. In warmer climates, the leaves may be evergreen.

Later on, I will pass into further element at the specifications this species desires to increase.

English Lavender is split into extraordinary kinds which are top notch options if you want to start developing lavender for profits. Here are some of them:

•Buena Vista

This is a tough-to-locate variety, I actually have to mention. It competencies adorable deep blue bi-colored flower spikes and a non-forestall bloom in most areas. In areas with a protracted growing season, you could assumeblooms consistent with year — overdue spring and fall.

This form of lavender is cherished with the aid of butterflies and bees, and may face up to the tough attempts of rabbits and deer. It is low upkeep, with a sturdy fragrance. It's a tremendous preference for mass plantings.

It calls for complete solar to broaden, further to low levels of water. Make sure to plant it in nicely-worn-out soils with low fertility. They thrive in warm climate and arid climates. These perennials are a exceptional preference for a drought-resistant garden, doing extremely good inside the drier additives of america much like the Great Plains, Intermountain West, and West Coast (which has a real Mediterranean climate (wet winters and dry summers). Hardy in zones five-9.

•Folgate

This English Lavender (L. Angustafolia) variety grows neat and tidy on a reasonably compact bush. Its flowers bloom a vibrant violet-blue that bees, butterflies, and florists certainly adore. Originally, this variety have end up superior in England inside the Nineteen Thirties. It prefers a miles less warmness weather.

It is a high-yielding plant with slight blue flora and moreover produces a immoderate extent of oil. If you make certain to maintain it warmth and beneath the daylight hours during wintry climate, it is even appropriate in big pots. This is why it is advocated for present day lavender growers and lovers. Hardy in zones 5-nine.

•Hidcote

Lavandula angustifolia 'Hidcote' is a decrease variety with a mounded increase dependancy. Hidcote is the hardiest of the English cultivars. It is a actual lavender that has been cultivated for its oil and dried plants.

It capabilities species like Hidcote Blue, Hidcote Pink, and Hidcote Superior. Blue is a dwarf plant, the Pink type has stunning pink plants, and the Superior type features dark purple plant life.

To develop Lavender Hidcote, it desires nicely-draining soil, ideally a piece sandy, and a sunny region. Hidcote can not tolerate humidity and couldn't perform nicely in areas which can be overly wet. In regions with higher humidity, ensure you can provide loads of air go with the flow. When planting, the crown of the plant need to relaxation sincerely on the ground of the soil. Hardy in zones five-nine.

•Thumbelina Leigh

Without a doubt, this is one of the favorites. It is quite small, with a compact, rounded dependancy. Its plants are a lovely, -tone darkish violet that appears fantastic in gardens. It does tremendous in boxes, and it blooms continuously from early to mid-summer season. It thrives in complete sun and does amazingly in dry to medium, mild sandy, alkaline, and properly-tired soils. Excellent drainage and proper air circulate are a want to if you need to expand this variety, specially in warm, humid climates.

•Munstead

Munstead is a smart, compact, dwarf (as a few would possibly say) shape of English lavender that consists of loads of deep purple-blue flora. Along with Hidcote, it's far one of the best sorts available that has stood the check of time.

Fully hardy, the fragrant stems of Lavender 'Munstead' may be cherished via all. They also may be reduce and dried for home made potpourri. It is truely a famous plant and can bloom twice a 12 months.

To grow this species, six to 8 hours of solar is quality, however some afternoon shade is first-class inside the hotter climates of the Southwest. Good air pass is crucial, specifically in regions of excessive humidity. Hardy in zones 5-9.

•Lavender X Intermedia

Let's soar out of the English Lavenders and check each different lavender species that's cherished by using growers all around the worldwide.

This species is really a skip amongst English Lavender and Portuguese Lavender and is stated with the resource of the decision lavandins. They are a chunk much less hardy however expand massive and will produce more flower spikes. They also usually tend to bloom later than extraordinary Lavandula species, blooming from July to September.

The varieties in this own family additionally produce large vegetation, and some produce a excessive focus of oil, this is exquisite to make extra products. The leaves, petals, and flowering tips can be eaten raw as a condiment or in stews, soups, and salads. However, as it has a quite strong flavor, ingesting in big portions isn't always encouraged.

They have large, gray-inexperienced leaves and are seemed for their rapid boom and sturdy perfume.

In terms of mild, it prefers entire sun (6 or extra hours of direct daytime an afternoon). It plays excellent in clay, loam, and sandy soils, with right drainage.

This species is separated into different sorts too, and they may be all nicely really worth finding out.

•Grosso

This lavender is probably one of the maximum planted across the area. It is extensively used for the producing of oil used cosmetically, which makes it relatively attractive for growers. It loves well-worn-out soil and warm climate with complete sun. It functions purple vegetation, long stems, and a increase dependancy that creates an almost exceptional 3-foot dome. Hardy in zones 5-9.

•Hidcote Giant

This species is quite well-known because of its colourful violet flower spikes, which is probably normally plump and large. It grows rather slowly, and buyers love to apply it in dried bundles, sachets, and potpourri. Hardy in zones 5-9.

•Impress Purple

This is called one of the most lovely lavender sorts. It is outstanding for clean bouquets. It capabilities abundant and strongly aromatic, rich darkish purple flower spikes growing on prolonged stems. It prospers in full solar and blooms from mid to past due summer season. Excellent drainage and air move are a must to domesticate this lavender. Hardy in zones five-9.

•Phenomenal

This French hybrid lavender is tremendous for its extremely good bloodless hardiness and tolerance to warmness and immoderate humidity. It grows right proper right into a stunning mounded form, with pink vegetation on tall stems in mid-summer time. Featuring vibrant violet-blue spikes, this plant is simple to boom and is drought tolerant as soon as it is set up. Reliably hardy in location 5.

•Provence.

This lavender sticks out from the rest as it's miles one of the extremely good lavenders for humid summers. Its fragrance is one of the most powerful (and that publicizes lots). Its huge flower spikes range as properly for their light purplish coloration. It is every so often known as "fat lavender" thinking about that its flower spikes can obtain up to 3 inches lengthy (eight cm).

•Seal

This shape of lavender is a real sight in the garden, with its violet-blue flower spikes developing on excessive leafless stems. It is a favourite for lavender bags even as you endure in mind that its easy, awesome fragrance can last more than one years.

•Spanish Lavender (Lavandula stoechas)

Most people, when they recall lavender, bear in mind the English and Lavender x Intermedia, but there are a couple extra who deserve the identical hobby. One of those is the Spanish Lavender. This species is favored through many growers due to the fact they're able to provide you with the equal fragrant flora, but they will be plenty greater tolerant to hot climates. In fact, it doesn't require any cold to deliver flora.

It is local to the hot climates of the Mediterranean, and it's pretty hardy in place 8. If you live in a place that doesn't depend with the four seasons or is truly warmer than the humid English climate, for example, this form of lavender is a superb desire for you.

At first look, Spanish lavender is pretty much like differing types, developing in small shrubs that make top notch low hedges or bed borders. They have the identical silvery green leaves, but what simply gadgets them aside from the rest is how they flower.

The pinnacle of every flowering stem grows big, upright bracts that resemble rabbit ears. Flowers may be crimson or crimson, depending at the cultivar. They have big, pine-cone-fashioned petals on the pinnacle.

They supply a smooth eucalyptus perfume. Popular as focal factors in courtyards and small-location gardens, Spanish lavenders take well to boxes and stylized pruning. It can amplify up to 60 cm (2 toes) tall mounds.

There are one-of-a-kind cultivars on this family, and they all have their non-public extremely good feature.

•Ann's Purple: they are massive than others and could broaden about 30 inches (seventy six cm.) all round.

•Purple Ribbon: produces dark pink vegetation, and it's far a touch bit bloodless hardier than exceptional cultivars.

•Kew Red: this cultivar is one of the few to offer adorable purple plants in a darkish raspberry color that catches the attention.

•Winter Bees: this one will start blooming earlier than one among a kind types. It starts offevolved in late winter in warmer climates.

•Lutsko's Dwarf: as its name implies, this dwarf cultivar grows out to approximately 12 inches (31 cm.) and makes a high-quality desire for concern growing.

To increase Spanish Lavender, find out an opening with complete solar or remember growing them in boxes; those plant life take nicely to pots. Make sure the soil is moderate and drains properly. Your Spanish lavender will no longer want some of water and will tolerate droughts remarkably well.

•French Lavender (Lavandula dentata)

Considering the decision, you likely have a superb concept of who modified into the primary to cultivate this type of lavender.

It is regularly in assessment to the English lavender, but there are a few crucial variations a number of the two to recollect before figuring out which to develop. French lavender is huge and will grow from about 2 to three toes (sixty one-91 cm.) tall and big, whilst English lavender stays lots smaller and greater compact, in spite of the fact that it may grow up to 2 ft (61 cm.).

Well-appropriate to milder climates without the scare of harsh winters, French lavenders are decorative plants known for their needle-like, toothed leaves (as a result their Latin call – dentata). And I intended the problem approximately harsh winters; French lavender is simplest hardy thru approximately zone eight and acquired't tolerate bloodless climates.

The plant life on those vegetation are similar in length to the English lavender, but they very last a whole lot longer on French lavender. This variety has one of the longest bloom instances, beginning in spring and continuing to provide vegetation at some point of the summer season.

Their perfume is lighter than the English lavender, so if you are looking for the real "lavender heady scent," French lavender might not be your splendid desire.

These plant life do well in rapid-draining packing containers and rock gardens and add a extraordinary dose of beauty even as lining walkways and get entry to paths. They pick complete solar and gritty soil.

Chapter 10: Growing Lavender

It makes me sad to say it, however not absolutely everyone can develop lavender. This is a species that is pretty practical to weather, air pass, drainage, humidity, and further. For starters, you have to have the right weather and the right soil, in spite of the fact that you could amend the soil to make it appropriate for developing lavender.

New hybrids are right at adapting to less warm and moister climates, however commonly, those flora live genuine to their Mediterranean roots. The Provence region of France is in which most lavender is grown commercially, due to the fact the weather is right for lavender manufacturing. It counts with moderate winters and warmth sunny summers, making it clean for lavender to develop.

Lavender grown in areas with excessive humidity often have problems with fungal ailments, however this will be avoided (or corrected) with the useful resource of through wider spacing a number of the vegetation to decorate the air flow.

It's real that lavender has been grown in maximum regions of the U.S, however some microclimates may be beneficial, like residing close to a big body of water.

If you're now not first-rate if your area is appropriate for growing lavender, ensure to check together along with your nearby agricultural extension agent or a close-by nursery or lawn center. They will assist you figure out if you are perfect to head.

What do you want?

•Soil:

In well-known terms, lavender prospers an awful lot better in a specific type of soil. It ought to have nicely-tired soil with a pH of 6 to eight. The pH is most effective a way of representing the soil's levels of acidity or alkalinity. Lavender prefers the unbiased range, with 6.Five to 7.Five being quite lots exceptional.

Why is that this important? Certain flora can best get admission to the soil's vitamins if the pH is inner a positive variety. And now not even extra plant meals or fertilizer will assist if your soil lies out of doors of a plant's brilliant pH range.

Testing your soil's pH have to be a venture to be added to your fall lawn checklist. That manner, you can repair the soil earlier than wintry weather or first element in the spring earlier than you plant. This is likewise an fantastic time to phrase any weeds which could have grown inside the direction of the summer season, which also can provide you with clues approximately your soil pH. For instance, dandelions and wild strawberries develop in acidic soil, on the identical time as chickweed, Queen Anne's lace, and chicory need alkaline soil.

How must you realize this price, you ask? You can check your soil with a easy pH tester found at maximum lawn facilities. If your soil tests alkaline, you may add sulfur to decrease the pH. If it is too acidic, upload lime to raise the pH.

If you'd like to recall a extra home made answer, you may get a tough estimate of your soil's variety the usage of kitchen factors. Here's how.

1.Take your pattern: using a shovel or your hand, dig 4 to six inches (10 to 15cm) below the ground of your garden.

2.Clean it up: Remove stones, sticks, and different particles from the pattern. Also, ensure to interrupt aside any huge clumps.

three. Add water to the soil: placed the soil in a glass situation and add sufficient water to show it into dirt.

four.Add half of Cup of Vinegar, and stir barely: If the soil fizzes, foams, or bubbles, your soil is alkaline.

five. If no effervescent occurs, repeat the machine (without which include the Vinegar)

6.This time, upload 1/2 cup of Baking Soda, and stir slightly: If the soil fizzes, foams, or bubbles, your soil is acidic.

Of course, you'll probably although need to head get the pH tester definitely to get the appropriate rate and accommodate therefore.

Apart from having the right pH, lavender does best in sandy loam soil that offers perfect drainage. If the soil will become saturated with water, as frequently takes vicinity in clay soils or a excessive-water table, it'll result in the roots rotting and killing the flora.

oIrrigation:

As I cited within the ultimate monetary disaster, maximum lavender species are pretty drought-tolerant that require fewer waters than maximum flora to do properly. However, younger lavender plants want ordinary watering till their root structures are nicely established. Besides, watering after harvest can assist produce extra flower stems for the following season.

Make superb to apply a drip irrigation approach, because it reduces the risk of fungal ailments and uses a tremendous deal lots much less water than overhead watering.

A extraordinary manual to conform with is to water a few times every week after planting until your flowers are set up. Mature flowers ought to be watered everyto 3 weeks until buds form, then a few instances weekly till harvest.

Make fantastic to check your weather carefully. If you've got were given non-forestall rainfall, permit Mother Nature do its element.

Once lavender plant life are mature, which takes spherical three years, they will be effective for about ten to 15 years. Make positive to recall that the extra mature, the a lot much less water it will need. Most commercial growers find watering absolutely as fast as eachweeks in the course of the brand new summer time months is pretty a wonderful deal right.

•Fertilizer

This is one of the belongings you acquired't have to fear approximately too much as soon as your lavender vegetation are nicely installation. Lavender does outstanding with almost no fertilizer at that point. However, even as they may be first planted, nitrogen fertilizer can help deliver them a lift. After the primary yr, it has a bent to stimulate leaf production in region of flower and oil manufacturing.

If with the beneficial resource of any threat you select out to fertilize your lavender plant life once they had been mounted, I may want to recommend and occasional nitrogen fertilizer. But in cutting-edge-day phrases, installed lavender acquired't really need it.

Growing for Profit

Growing lavender for income may be a pleasing and profitable enjoy, as long as you apprehend what your vegetation want if you want to develop robust.

If you are making plans to increase for a commercial enterprise business enterprise possibility, I might endorse installing trial regions or about a hundred fifty-two hundred plants each, with handpicked cultivars to help you outline the extraordinary and yields earlier than going with the large scale.

There are early, mid, and past due season flowering sorts to be had, so have a study your climate very well and determine as a manner to be the superb planting time for you. As mentioned, you can want moderate soil with first rate drainage, and if you are able to providing it with a slope to the south or southwest, they may thanks. Note that lavender types are vulnerable to frost damage.

How plenty cash can a lavender enterprise make, approximately?

The amount of cash you're making will rely upon the dimensions of your operation and the way well you're taking gain of various advertising and marketing and advertising and marketing techniques.

Some small growers choose to live with some dozen flowers of their out of doors and are happy to make a few hundred dollars. Bigger operations on acreage can convey in masses of heaps, particularly if in addition they produce and promote price-brought merchandise.

Fresh lavender bouquets are a terrific, profitable manner to promote lavender. Most growers promote direct to the retail public, each from their lawn or at the farmer's marketplace.

Lavender bunches can sell for approximately $6 every at neighborhood markets. A 20' x 20' developing area can produce around three hundred bunches each yr, nicely actually well worth $1,800. Larger plots are even greater worthwhile. A region-acre can produce about three,000 bunches, without a doubt well worth $18,000.

Other lavender merchandise, which encompass creams and soaps, convey 500% or more markups from the price of the primary materials.

Many growers consciousness on selling artisanal and domestic made lavender merchandise through social media and at their community shops, that could boom your earnings and reap a fair wider target audience this is interested in different topics aside from bouquets.

For most small lavender growers, another relevant way to promote is with the aid of manner of attending the Saturday market. From easy lessen lavender bouquets to dried buds, lavender oil, and the severa price-added lavender products you could make and sell. Best of all, promoting direct on the marketplace lets in you to lessen out the middleman and get hold of complete retail costs on your merchandise.

There are multiple lavender merchandise you may installation and promote. Satchel, as an example, may be used anywhere the air desires freshening or deodorizing, which consist of drawers, closets, even in smelly footwear. Most sachet earnings come from repeat customers who love the perfume of lavender. Sachets are also supplied to neighborhood shops.

A pretty unusual lavender product is dream pillows (Dream what?). As we mentioned earlier, lavender is notion for its calming impact. Putting it in a pillow makes enjoy to assist inspire restful sleep and is one of the maximum profitable rate-delivered lavender products. Medical research have even decided that lavender can assist calm kids with ADHD. One enterprising lavender grower has created a line of animal-themed dream pillows for kids, grossing over 1,000,000 bucks every 365 days from her pillows.

One fantastic product to feature for your repertoire is lavender flea repellant. Many business flea repellants use effective chemical substances that might have poisonous facet consequences. Lavender is an all-herbal flea manipulate that now not best repels fleas; it could even help you're making pets scent better. The traditional markup over the fee of substances is spherical 500% to 800%.

Later on, I'll tell you about all of the one-of-a-kind delivered-rate products that you could make with lavender, much like the well-known lavender cleaning soap or perhaps dried bouquets!

Propagation

To ensure that lavender flowers are regular in outstanding, oil manufacturing, shade, and length, almost all agency growers propagate lavender from cuttings in region of from seed. Producing lavender from cuttings guarantees the new flora can be real clones of the figure vegetation.

One of the easiest methods to propagate lavender is thru taking cuttings - and the great time to take cuttings is after flowering. The component is that `pruning lavender as a minimum two times a yr is crucial to make certain it remains healthy, so there's no harm in taking the offcuts from habitual pruning to create even greater first-rate lavender flora.

Keep in thoughts that it's going to take round a three hundred and sixty five days for the propagated lavender to emerge as massive enough to plant.

What you'll want to propagate lavender

•Sharp secateurs

•Rooting hormone powder like (Hormodin® or Dip-N-Gro®)

•Small pot

•Seed raising aggregate

•Pencil

Here's a way to propagate lavender from cuttings step by step.

 Step 1: Choose the proper candidate. Make remarkable to choose non-flowering shoots that have a woody base however with a moderate, green tip. Gently pull a 10cm shoot to the component and strip it a long way from the number one plant, ensuring it has a heel (a strip of bark) connected. Trim with secateurs.

 Step 2: Give it energy. Remove leaves at the lowest of the decreasing and dip the reducing into the rooting hormone powder that's appropriate for semi-hardwood cuttings.

 Step 3: Plant it. Fill the pot with the seed-raising combination, which include half peat moss and half perlite, vermiculite, or sand. A properly-worn-out blend enables save you fungal illnesses. Using the pencil, poke a shallow hole in the pinnacle of the mixture and insert the reducing. Repeat for each decreasing. Firm the cuttings into the aggregate together with your palms and water the aggregate. In bloodless regions, make certain to cover the pot with a clean plastic bag secured across the rim with an elastic band.

Place the pots on a warm windowsill and water on the identical time as dry – take care not to overwater, as this will cause cuttings to rot.

And that's it. You without a doubt must wait a one year (approximately) to start seeing your stunning cuttings trade into vegetation.

Chapter 11: Preparing For Cultivation

Now we're on the point of get into the great subjects. By now,you need to have nicely-worn-out soil with the right pH. But there are a pair extra factors that you want to recollect in advance than going out and getting geared up the soil to your lavender.

•Buying lavender to get commenced out out

To get your first lavender plant to assemble your destiny empire, you want to recognize that flowers are extensively to be had in the route of spring and summer season in garden facilities and on line. They are commonly offered in bins – 9cm (3½in) or large – organized for planting. Lavender also can be located as plug flora in spring thru a few mail-order providers. This is a cheaper manner to buy, mainly in large portions, but the preference of cultivars is confined.

Also, these tiny vegetation want to be looked after carefully for numerous months before they may be massive sufficient to plant into their final function.

•Site Preparation

When you're geared up to start making equipped the ground, keep in thoughts that Lavender is splendid planted in April or May in maximum areas, because the soil truely warms up, and it's the time while smooth flowers come to be available in garden centers. Lavender need to by no means be planted in wintry weather at the same time as more youthful plant life are prone to rotting in cold, wet soils.

Now, let's dive into a few precious soil tips.

According to Curtis Smith, Ph.D., the best soil consists of forty five% mineral (sand, silt, and clay), 5% natural count number, and 50% pore location. The pore area is what offers plants oxygen and water essential for growth.

Depth is each exceptional thing to maintain in thoughts. The intensity of right soil coaching can decide wintry weather survival, nutrient availability, and potential for damage by means of way of spring frosts and summer season droughts. All of these right now impact the growth of floral stems, buds, and crucial oil of your lavender vegetation. Shallow soil schooling manner you could get shallow roots, and plant life with shallow roots are more touchy to summer time droughts.

Here are a few sizeable guidelines to bear in mind:

Avoid pulverizing the soil, as this will create micropores inside the soil, which make it an awful lot more hard for the plant to drag water from them.

Use amendments which incorporates compost, wood chips, bark mulch, and extra. These will decompose and assist decorate your soil's shape, making it an entire lot more exquisite and nutrient-crammed in your lavender.

Planting Lavender on raised beds permits you increase the drainage of excessive quantities of water, as it is able to display up at some degree inside the rainy season. Planting on raised beds allow you to lessen the opportunities of root rot.

Use herbal rely range that is in reality coarse. Also, try to keep away from using natural recollect this is excessive in soluble salts. For example, manures comprise salts and may boom the soil salt diploma even as used as mulch or soil amendments. Lavender will tolerate a rather immoderate salt degree, but it is able to have an impact on its health and yield.

 Never artwork the soil even as it's miles wet. This destroys the shape it creates with the amendments and natural depend amount.

Apart from getting prepared the soil, ensure you cope with any weeds near the planting spot. They will compete with lavender flowers in terms of region, get right of access to to sunlight hours, water, and nutrients. The presence of weeds could have a terrible effect on the quantity of glowing plant cloth harvested as well as on the best of critical oil.

The first measures closer to weeds want to be excited thru the number one tillage, nicely in advance than planting the number one lavenders. That manner, you may ensure your plant life received't be competing with weeds for boom, slowing down their system.

Planting Lavender

Now which you recognize all that lavender desires so you can develop (the proper climate, right soil, correct watering, and more), it's time to learn how to clearly plant your stunning seedlings or cuttings. Let's dive right right into a step-with the resource of manner of-step manual on a way to plant lavender at your private home or each other region of your choosing.

For nice outcomes and most enjoyment, you may location lavender close to doors, domestic domestic home windows, and paths to get the entire blessings of the perfume. Also, that manner you may be capable of study because the bees and butterflies swarm sooner or later of it. While you sip your lavender-infused gin and tonic, of route.

The first step is choosing the right area. Find a sunny spot with the right soil conditions. As I've noted earlier than, it received't do properly in shady, damp, or cold situations.

Try to get your lavender planted as short as you can once you've had been given it home. If you're doing it with a superior plant, make certain it's far healthful and has superior an super root device.

While it's far viable to boom lavender from seeds, I wouldn't advise it, due to the fact the seeds require scarification and chilling and may take nearly a month to germinate.

Prepare a planting hole that's two instances as deep andtimes as huge as the concept ball of your lavender plant. If the roots are clinging to the edges of the pot, you could "tough up" the roots to encourage outward increase.

Lightly prune your lavender in advance than planting. This will deliver the plant form, ensure true air movement via the stems, inspire new boom, and prevent the center of the stems from turning into woody, that is a commonplace trouble with lavender.

Some growers recommend shaking the roots a piece to cast off the soil greater earlier than planting.

Plant your lavender with the pinnacle of the idea ball, despite the soil line. Fill in any more area spherical and above the lavender's roots with soil, lightly patting it into location around the base of the stems.

If you're planting multiple lavender plant, depart about 36 inches (91.Four cm) among every plant. This will assure right air stream and allow the lavender place to grow.

If you are thinking about fertilizing your plants to offer them a moderate push, an outstanding time to do it's far after the first watering. Allow the soil to dry, and then practice the fertilizer.

Lavender grown in containers can also need a hint extra hobby, mainly inside the top of summer season, because the soil will dry out brief in warm weather. In winter, keep packing containers pretty dry and stand them on their ft to help drainage.

After planting your lavender plant life, preserving them and troubleshooting is the subsequent challenge.

How to maintain lavender?

•Problems with overwatering:

Apart from watering, lavender calls for some other cares so that it will help them maintain in a lovable shape and yield the incredible outcomes feasible. Here are a few more watering tips to consider:

•To gather the right stage of watering, make sure that the soil dries amongst each session.

•If you are developing lavender in a Northern climate, you will water the plant very sparingly until the summer time, while temperatures can skyrocket and dry out the soil. You will then need to begin watering the plant every 7 to 10 days.

•If you're developing lavender in a pot, ensure the pot has incredible drainage to prevent water from pooling at the bottom.

The maximum apparent signs and symptoms and symptoms and signs and symptoms that your lavender is death from overwatering are signs and symptoms of strain, which includes a drooping or wilting look and a browning of the foliage.

It's a tremendous not unusual mistake to don't forget the drooping appearance and brown foliage as an underwatered plant, and loads of growers try and treatment the problem with greater water. Consequently, the premise rot turns into worse, and the plant dies fast.

If you water your lavender as regularly as one-of-a-kind plants to your garden, you may in the long run kill the plant.

If via any hazard, your lavender plant life ever show the signs and signs and symptoms of being overwatered, then you may want to prevent watering them for at least 3 weeks. If feasible, protect them from rainfall.

This will supply the soil a risk to drain and the roots the opportunity to dry out and get over root rot.

You will want to prune any affected foliage absolutely below wherein it's far brown with a sterile pair of pruners.

After 3 weeks with out water, the plant want to appearance a good deal more healthful, and you can resume a regular watering time desk.

For in recent times planted lavenders, the watering need to be spherical as quickly as in keeping with week for the primary 4 weeks.

For lavender inside the firstyears of increase, as soon as eachweeks if there may be little or no rainfall all through this era.

•Problems with daylight

In their nearby Mediterranean variety (Italy, Southern France, and Spain), lavenders experience complete sun all day. You do not want a Mediterranean climate to increase lavenders, however you need to ensure that they're inside the sunniest place of your garden.

The amount of sun your lavender receives will right now have an impact on the quantity of flora, oil, and aroma your lavender will produce.

Lavenders want as a minimum 6 hours of direct sunlight hours per day during the spring and summer season months to expand successfully. If they accumulate much less than 6 hours of daylight, lavenders will gift disappointing growth, a loss of colour in the leaves, terrible fragrance, and will probably die.

If your lavender ever presentations horrible boom and a lack of flowers, then make sure to preserve in mind the amount of daylight hours it gets. If it's an lousy lot an awful lot much less than indicated, drift the plant to a sunnier location for your garden.

•Pruning

Lavender desires pruning in the course of the primary years at the same time as it's miles despite the fact that developing. If they aren't pruned within the direction of that length, the stems ought to turn out to be woody, resulting in fewer stems and plants.

The number one golden rule with reference to pruning lavenders is to great lessen again into inexperienced foliage and never reduce yet again into the woody boom. Cutting another time into the woody boom will cause the lavender splitting and forming a awful form or even dying from marvel.

For terrific outcomes, you want to lessen decrease once more the spent flowers in overdue summer time and prune lavenders inside the spring.

Pruning must take location whilst new leaves begin to grow at the bottom, that is generally around early spring. You can prune as plenty as a 3rd of growth from the pinnacle with the purpose of shaping the lavender so it continues a rounded form. This prevents the plant from splitting.

A visual manual on a manner to prune is the incredible choice in case you don't recognize a way to do it, so you can check out this first-rate video on YouTube through Garden Time TV.

•Excess of nitrogen inside the soil

If the foliage of your lavender ever turns yellow and likely has a leggy look, this could be an illustration of a in addition of nitrogen for your soil.

In order to increase, flora need nitrogen, phosphorous, and potassium. A higher degree of nitrogen in the soil may be an extraordinary problem for some plants, but not for lavender.

Lavenders thrive on nearly neglectful care and bring more vegetation in soil that is medium to low fertility.

Established lavenders will not need any greater fertilizer. This advice is confirmed through the English Royal Horticultural Society.

The answer is based upon on what the supply of the immoderate nitrogen is.

If you operate fertilizers in your lavenders, you then definately want to end the use of them proper away. The plants ought to show a higher face subsequent season.

If the flora didn't accumulate any fertilizer and the lavender is yellow, then it's far viable that the soil is sincerely too excessive in vitamins (wonderful, soil may be too top for flora).

In this case, you could flow into your flowers to a pot with 70% potting soil and 30% sand or grit. You also can dig up the lavender and upload loads of sand and/or grit to the soil in advance than replanting.

Chapter 12: Harvesting Lavender

This is the fantastic thing for all lavender growers even as all the difficult artwork in the long run will pay off, and you get to pick out up your stunning plants.

When you're without a doubt starting out, it's most effective regular to marvel approximately at the same time as the high-quality time could be to attain. Will it's far in some months or in 3 years? If you ask any professional, they'll possibly let you understand that it usually takes spherical 3 years for flowers grown from seed to attain their top, yielding a miles greater harvest.

However, developing lavender from cuttings will purpose them to bloom faster and reflect the arrival of the specific plant. With those, you may generally harvest inner some months, even though this will vary relying on the species.

•How do you recognize they'll be organized to be harvested?

#1   #2

By looking your lavender plant life enlarge and remodel in some unspecified time inside the future of the season, you'll growth a enjoy as to even as the time is proper for harvesting. As the climate warms within the spring (and after a nice spring pruning), lavender plant life begin to deliver up stems and tightly closed spikes.

Buds form on the spikes and develop, in the end changing from green to a greenish hue of lavender (or white, or blue, or red...). Then, the vegetation (corolla) emerge from the buds – The lavender is blooming.

If the plant is displaying buds which might be genuinely closed, and that they however have a translucent, greenish hue, then it's too early to achieve.

A picture containing flower, plantDescription automatically generated

Look at the difference a number of thestems above. One is in complete bloom (even as your lavender seems like this, you'll realise it's time to collect). The one at the left has closed buds and no flowers.

Generally, you'll want to reap a stem of lavender whilst approximately half of of the buds are in bloom.

The nice time an wonderful way to reap will change counting on the motive you have for it. This is why I advocate that, earlier than harvesting, be organized. Think about what you want to sell for your capability clients and why.

For instance, even as harvesting lavender for critical oil distillation, wait till 50%-100% of the buds are blooming. Or whilst harvesting lavender for dried buds to use in potpourri, sachets, or culinary makes use of, harvest even as 25%-50% of the buds are blooming.

•How to harvest

Now which you have an idea of what you're searching out, you could grasp your scissors – or your hand sickle – and get available. The best way to analyze is through doing. The greater time you spend around your lavender plants, the greater you'll start to select up on their subtleties.

Cut low so that you get prolonged stems, however don't reduce into the woody base of the plant, as this could stunt new growth next yr. That may be irreversible damage. You can check for where the "green boom" ends and decrease only a few inches faraway from the "woody increase."

You can reduce a gaggle of stalks at a time if your tool is able to lowering via them. Clean cuts as adversarial to ripping the plant's stalks.

Bring rubber bands so you should make bundles of lavender as you harvest, which you can experience glowing or hold upside-down in a fab, darkish region to dry. You can pull off some of the leaves as you harvest to rush up the drying method.

And of path, you constantly need to acquire inside the early morning, in advance than the warmth arrives. Lavender loses its oil to the warmth of the day, so harvesting in the cool of the morning (in advance than 10 am) manner you'll harvest lavender with better oil content fabric material.

Also, you don't want your lavender to be moist from dew or maybe from a middle of the night water sprinkler. It will take longer to dry, and you may hazard mold.

So, you've controlled to efficiently harvest your lavender. What's next?

•Drying lavender

Drying your lavender bundles is the subsequent step of this method. Dried lavender has an fantastic amount of makes use of, and they're pretty smooth to dry. It's used to flavor ingredients, make lavender tea, make lavender cleaning soap, upload to potpourris, and extra.

Here's how you may dry your lavender (to the factor wherein they sense like potato chips) after harvesting them at domestic.

The simplest, most commonplace approach to dry lavender begins offevolved via tying your lavender bundles with the rubber band and finding a darkish, dust-free area wherein to hang them the wrong way up.

Typically the bundles are hung in a groovy room. An attic room should artwork very well and maintain your herb bounty out of sight for a while. You can set up nails or hooks along the partitions or ceiling beams and hang every package deal allowing air movement amongst them.

After some time, your lavender bundles will feel as crispy as a chip. That's whilst you'll need to prepare for storage (we'll get there in a 2nd).

There are loads of drying strategies you may undergo in mind, despite the fact that some of them will rely on the shape of climate in your place.

•Air Drying:

Outdoor air drying, together with in an outbuilding, is a commonplace area for drying lavender in dry climates. However, many new cultivars of lavender had been tailored to humid environments, and air drying in humidity isn't always as smooth. In that case, you'd need to dry your lavender indoors.

•Dehydrators:

Dehydrators paintings well for lavender if you maintain the temperature under one hundred fifteen°F. Snip stems with lavender blossoms which might be surely open. Clip off discolored or shrunken leaves and check for insects that might be hiding inner blossoms or leaves. Do not wash the lavender. Just place the cut stalks in a unmarried layer at the dehydrator trays.

Set your dehydrator on an herbal placing and allow it paintings for about 2 hours. If the blossoms, stems, and leaves are brittle and paper, it's miles completed. If it although feels wet, or extra like a tortilla in choice to a chip, depart it for some different hour to dry.

•Fans

If you don't have a dehydrator or want to dry huge quantities, the usage of a warmth, blanketed location and an electric powered fan can help. Don't cause the fan without delay at the lavender, or it will dissipate the fragrance and dry the lavender too fast. You don't want that.

Instead, purpose the fan, so air circulates below and across the lavender to move humidity and maintain moisture from settling. Alternatively, using the fan as a vent (through putting the fan subsequent to the lavender but blowing a ways from it) can assist.

•Ovens

Lavender may be dried in a grew to grow to be-off oven. The first-class conditions for drying lavender in the oven are a temperature of round ninety°F, with very low humidity, particularly the buds. Gas-lit ovens are suitable for that.

In an electric powered oven, you may set it at one hundred°F or its lowest setting. Warning! Don't use the "warm" setting; that's too warm.

The drawback of the oven technique is which you lay the lavender down. That's brilliant for the blossoms, but if you attempt to dry long-stemmed lavender bunches, they dry unevenly, and some flowers get beaten.

After drying, comes the a laugh detail. This is in which you'll eventually begin making prepared the charge-introduced products that you may set up and promote to customers. Make positive to notice precisely a way to make each of those merchandise, so you can later decide the species you may plant, the aftercare, and the drying method on the manner to paintings wonderful.

Let's dive into it.

Chapter 13: Products On The Way To Increase Lavender Fee

As I've referred to in advance than, developing your earnings with lavender is the precise element. By taking some extra steps, you could take your harvest from a fundamental herb to treasured merchandise that supply pinnacle dollar from clients and high-profits margins – in truth, growers around the arena have said they're capable of double, triple, or quadruple their income with rate-introduced products! Doesn't that sound extraordinary?

How can this be, you ask? Take a check how the market has modified currently. The herbal and private care company has grown to over $6 billion a one year and keeps to rise with each day this is going through using. More and more fitness-conscious clients pick out to go along with chemical-loose, natural substances in place of synthetic ones. And lavender is an outstanding desire for this.

A huge part of that money goes to small entrepreneurs who found their spot inside the marketplace with their home made, herbal merchandise that supply what their customers are inquiring for.

A suitable example of that is mothers who're tired of buying unreliable and non-obvious products for their babies, like little one butt balms for diaper rash. Plenty of them have located their solution through manner of developing their private salves and creams, the use of handiest natural factors that they may rely on.

"I want to recognize precisely what's inner it, and it's no longer always smooth to buy a very natural product." This is the general sentiment that propels them to find out their very own solution.

There are everyday troubles that need a deeper solution than what big agencies can provide. When a young toddler gives dozing problems at the same time as they'll be making their transition from a crib to a mattress, they might use the herbal calming homes of lavender to lessen the tension of children.

There's one entrepreneur, in particular, referred to as Lauren Rosenstadt, who placed her gold mine at the identical time as she heard a pal saying that her daughter had insomnia. She had the tremendous idea of creating a "dream pillow" for her the usage of lavender. It have end up in the form of a sheep, and she or he or he called it "Sydney Greensheep."

The next day, her friend called her in awe, "It's implausible. She slept like a little one."

Lauren didn't forestall there. She knew she had placed an brilliant place to discover, and so she began out making greater. She took them to a change show and provided over one hundred pillows. By the prevent of the 12 months, she had made half a million dollars. By now, her empire has grown.

On the About internet web page of her internet website online, you could find these great phrases, packed with reality and imaginative and prescient: "We've taken a cue from the centuries-vintage culture of the usage of herbs to help summon candy dreams--to create adorable sleep companions."

You can test out her cute pillows proper right here: https://www.Herbal-animals.Com/

And that is exquisite one fulfillment tale among many. Hundreds of marketers in one-of-a-type additives of the arena apprehend that there's a rising opportunity with lavender. They ought to make conscious products while not having excessive technology and the usage of not unusual sense of their advertising efforts.

You don't need a chemical diploma to combine some herbs and make lotion, soaps, and whatnot. Here, you may discover a few super thoughts that you may take and deliver them a private contact. Just ensure that a few element you do, you do it together with your complete coronary coronary coronary heart.

Having a business enterprise is not best approximately creating a residing however about supporting others. Keep that attention, and you'll be at the way to a hit entrepreneurship.

•Live lavender plants

When you keep in mind selling lavender, this one is probably the primary that involves mind (as may be the case for plenty exceptional ordinary plants).

You can continuously provide stay potted lavender flora at a close-by farmer's market to inspire growers and fanatics to plant lavender at home, too. They are pretty clean to sell or perhaps less tough to offer in case you begin them from cuttings. They grow speedy, and the charge of producing them is low. Just some cents for a pot and a bit of soil, and you are ready to move. Their profits margin is specially excessive.

Most growers discover that the 4 inches and 6 inches pot sizes promote the super. Also, you could play around with the product. You may also additionally want to promote them alongside a clean bunch of lavender flowers, truly so the client has an concept of what they may get with the resource of trying to find the stay plant.

•Fresh lavender bundles

This is one of the maximum well-known lavender products to sell. When people find out a stall promoting lavender, they'll routinely expect they may have the potential to shop for a bundle deal. The thing approximately bundles is that the timing desires to be proper. They ought to sell quickly after harvesting, so they won't lose any appealing property.

Make high exceptional you pick out a gap with masses of foot web web page visitors, like a avenue sincere or the network Saturday market. You can also try to sell bunches to shops, like florists and grocery shops. Also, presentation is essential too. They need to be quite and presentable in case you need them to sell. Use a plastic floral sleeve for every bunch, and tie them with a handsome ribbon.

•Dried lavender buds

Lavender buds are a growing product right now. Some humans like them for making their very non-public sachets, bath bags, and closet fresheners. They have even been used to throw like confetti at weddings. How crazy is that? They may be bought for definitely one in all a kind fees, counting on the size of the bag.

If you need nice dry lavender buds or petals, you could cast off the lavender stalks from the sparkling lavender, keeping the lavender heads.

Place your lavender heads in a layer at the bottom of a field protected with newspaper. Store this problem in a warm, dry area, and lightly shake each day to aerate. Once dried, rub the lavender heads to cut up off the character buds.

•Dried lavender bundles

Dried bundles may be offered to crafters, who typically generally tend to use the stems to create floral arrangements, wreaths, and precise products. Florists are each other normal purchaser who might also use those bundles all 365 days prolonged in floral arrangements. You can find out them on Etsy or Amazon, and the costs range depending on the grower. They appearance lovely alone, and they provide out a fragrance that's calming and soothing wherein positioned at some point of the house. For this motive, a couple of clients would really like to look this in your product repertoire.

•Sachets

I've been bringing up this product at some point of the whole e book. There's a cause why this is one of the satisfactory sold lavender merchandise to be had. There are dozens of makes use of for lavender sachets, so they will be steady sellers at Saturday markets to duplicate clients. Lavender sachets may be applied in dryers to provide garments a diffused lavender scent, as a bath bag, as a closet or drawer freshener, in stinky footwear, and anywhere the air desires freshening or deodorizing with out using synthetic chemical fertilizers.

They also are used internal pillowcases to encourage sleep after an prolonged day with a racing mind. Parents around the sector embody a sachet interior their children's storage packing containers or closets to keep away from the odor of plastic or timber.

They go away a number of room for creativity, too. If you do a brief searching for in places like Etsy, you'll see all of the considered one of a type techniques in which growers and crafters format sachet luggage and packages. They supply them a non-public contact that certainly makes a distinction.

Lavender dryer luggage are an smooth-to-make possibility to expensive chemical-crammed dryer sheets and are continually splendid-dealers on the Saturday market. To reason them to, just fill a bag with about a heaping tablespoon of lavender buds. You'll want to apply a choice with a robust perfume, like Grosso.

•Lavender cleaning soap

Like sachets, that is a well-known taken into consideration clearly considered one of lavender growers. You can play around with the scale and form of your cleaning cleansing cleaning soap. It's up for your creativity. They are an aromatic item to have in any toilet. Best of all, cleansing cleansing soap is a repeat dealer and moreover well-known for offers. Even beginners can produce a saleable bar of cleansing cleaning soap with the useful resource of the usage of a clean soften and pour base and a mold.

To make lavender cleaning soap, you may want: Goat's milk glycerin, lavender sprigs, lavender essential oil, cleaning cleaning soap colorant (non-obligatory), and a cleansing soap mould.

•Break up your cleaning soap base into smaller chunks and location proper right right into a bowl.

•Melt in your microwave at forty-second durations until the cleaning cleaning soap is fully melted.

•Add your cleansing cleaning soap colorant (when you have any) and important oils and stir nicely.

•Pour your cleansing cleansing cleaning soap into your mold. I suggest the usage of soft and flexible molds, so that you gained't need to do some component particular to the mildew. If you are the use of a bread pan or a non-flexible mold, you could want to spray a piece little bit of cooking spray to assist the cleaning cleansing cleaning soap ease out of the mould even because it has hardened.

•You can upload the lavender springs to the front of the cleansing soap and permit it dry with the relaxation of the cleaning cleaning soap.

Make exceptional to create cute packaging. Let your creativity shine. You can actually have little playing cards designed to consist of them in every purchase.

•Lavender candles

Similar to soaps, lavender candles are a amazing addition for your merchandise considering that they'll be an ever-well-known fragrant stress reliever that may be produced in an brilliant sort of shapes, sizes, and hues. They can be embellished beautifully, and you may provide them a personal touch. In truth, putting them in Mason Jars turns them right right into a beautiful present for weddings, birthdays, or to enhance your home. So, how do you're making lavender candles?

Here's what you'll need:

•Two glass mason jars with big mouth openings and smooth ground: one 8oz and a few exceptional 12 oz.. (The smaller jar should fit into the larger one with a small hollow among surfaces.)

•Dried lavender stems

•Candle Wax – white soy wax or white paraffin wax, 2 cups

•Candle wick with clip and decal, medium size

•Double boiler

•Wood skewer

•Scissors

•Lavender crucial oil

•Twine (optionally to be had)

Ensure your glass mason jars are smooth in advance than the use of them.

Slowly soften the candle wax the usage of a double boiler. Make sure you don't overheat it to avoid burning the wax. Once it's melted, do away with it from the warmth. In the interim, stick the clip of the wick to the center backside of the small Mason jar.

Add 15 drops of lavender vital oil to the melted wax and stir. (TIP: You can constantly upload greater oils if you'd need to get a more potent fragrance).

Slowly pour the wax into the prepared jar.

Wrap the prolonged end of the wick spherical a skewer and relaxation the skewer all through the top of the jar. This will hold the wick centered for your jar as you pour the rest of the wax.

Leave the wax to absolutely cool and set. Once the wax has set, reduce the extra wick ¼" from the pinnacle of the candle.

This is in which you could get innovative. Gently vicinity the small jar into the larger jar. Working one stem at a time, accommodate and insert the dried lavender stems inside the area some of thejars.

This ornament will make your candle look colorful and sensitive.

Continue running all of the way during the jar till you reach the quality surrender stop result. Wrap some wire throughout the neck of the jar with some stems of lavender if you need. It offers it a rustic contact this is going exquisite with the delicacy of the stems.

You can mess around with this. You can also want to color the mason jars or stick high-quality substances to them. This is a product that allows you to get innovative and supply every candle your personal contact. You can sell them at neighborhood markets, grocery shops, or perhaps online. Platforms like Etsy are a tremendous desire for domestic made, artisanal products like this.

•Aromatherapy oil

I've been bringing up this at a few level in the complete e-book, however lavender oil has a unique soothing effect on humans at the same time as it is inhaled. It is regular for masses enterprise growers to encompass a bit, private-sized bottle of oil with distinctive non-public care merchandise. So, via now, you need to be thinking: how can I make lavender oil at home? Allow me to give an purpose of.

Step 1: Cut off the sprigs from the lavender.

The very first step in making lavender oil is to lessen the sprigs from the clean lavender. Each section have to be 6 inches or shorter. It's essential to check that you can employ the leaves and the stems within the oil-making approach, however I wouldn't suggest you to use the thick stems which is probably close to the lowest.

Step 2: Dry the lavender

After reducing the sprigs, follow the stairs to allow your lavender dry thoroughly. Drying the lavender flowers complements their aroma and reduces the opportunities of the oil getting dry.

Step 3: Cut the glowing lavender

Once your lavender is dried, the following element which you want to do is to begin crumbling the lavender flora. Once you finish crumbling them, location them smartly in a jar.

Before you area the lessen lavender within the jar, make sure to clean the jar thoroughly and dry it completely. The mixing of oil and water may also cause infusion, and you don't need that to reveal up in this case.

Step 4: Pour oil over the flora

After placing the sparkling lavender in a easy jar, it's time to sprinkle a few nearly heady scent-loads much less oil collectively with almond oil or olive oil at the flowers. Make excellent which you do it in a manner that the oil absolutely covers the vegetation.

Tip: Smell the oil on my own in advance than inclusive of it to the jar, truly in case you don't love it. If it's too robust, you can need to try unique options.

Step 5: Soak the lavender

After pouring the oil at the plant life, take the jar and location it in a sunny region. You will want to allow the flora soak for as long as you may. This is what is going to carry out the aroma from the plant life and infuse it within the critical oil. The least amount of time that I may endorse for this step is forty eight hours, however in case you aren't in a rush, you could permit them to soak for as long as six weeks.

The difference in aroma most of thetimeframes is surprising.

Step 6 (and a 1/2 of): Heat the oil

I get it, once in a while you won't have the time to wait 2 days or perhaps 6 weeks to have your vital oil equipped, so there's every other alternative for you. If you don't have a good deal time to soak the flora in daylight, you can attempt out this opportunity method, which includes heating the oil and flower mixture both in a boiler or a crockpot at a excessive temperature.

Make positive to have a cooking thermometer on the equal time as acting this step, as too much warmness may also ruin the aroma of the plants. If that is your first time making oil this manner, it would take some attempts so as to get the temperature right.

Step 7: Strain the oil

After heating the combination, you will need to strain it. You can do it via putting a cheesecloth or muslin over a bowl and then pouring the aggregate into it. After the straining approach, ensure you don't throw the very last flowers and exclusive small portions of lavender inside the bin. Instead, use them as compost in your gardens. The relaxation of your flowers will thank you.

Step 8 (only if you want to be the super provider in the market): Make the oil more potent

If you'd need to make your oil more potent and improve the fantastic of your product, repeat the method defined as masses as you may. Do it five, 6, 7 instances in a row if you have to. Just make certain to feature more lavender portions and attempt different things. Test the difference amongst heating the oil, leaving it to soak for forty eight hours, and leaving it for over a month.

The manner to achievement is paved on repetition and variation, a very good way to make an appropriate vital oil, you need to dive into the practice.

After your oil is prepared, you may divide it into small bottles to accompany awesome lavender products, or into large bottles to be bought on my own.

Either manner, that is one of the top notch lavender products you could sell, and as you may see, it doesn't take an excessive amount of to make it.

You can tell your customer that further to its calming effect, lavender oil is antibacterial and often used to cope with cuts, scrapes, and bruises, similarly to extraordinary pores and pores and pores and skin issues like pimples and athlete's foot.

Lavender oil, like most vital oils, want to be saved in hermetic, dark glass boxes in a groovy area away from direct daytime.

•Dream pillows

I've stated this some times by using way of manner of now, so it's approximately time I certainly informed you what they are and the way to reason them to. Dream pillows are easy to make and feature a immoderate income margin, as the value of the substances is so low.

In brief, a 'dream pillow' is a small pillow packed with relaxing herbs like catnip, lavender, and mugwort. A dream pillow is supposed to prevent nightmares and sell restful sleep, taking advantage of the herbs' herbal fun effects.

This is one of the satisfactory products to sell, and truly each person who has ever produced it's miles privy to they will be a ordinary supply of profits. Think about it. We all understand a person who has hassle napping.

For the maximum factor, I sleep nicely, however I even have such a number of friends and own family people who don't. And many growers spherical the place could believe me, announcing that what they want is a dream pillow.

Aromatherapy is the buzzword implemented in contemporary-day times to provide an reason behind what certain scents will do for our mind, frame, and soul. So, how will you build your pillow?

There are many recipes you could have a take a look at, and maximum of them consist of lavender as one of the most essential protagonists. If you choice, you can purchase some additional herbs to complement your lavender, and also you'd be presenting a miles more effective dream pillow in your clients.

For example, this is an awesome recipe for restful sleep: half of of cup lavender, half of of of cup mugwort, and 1/2 of cup candy hops. Blend all substances and simplest usetablespoons ordinary with pillow. It's greater than sufficient.

So, how do you assemble the pillow itself?

You will very in all likelihood have a number of pieces of scrap cloth spherical. If you don't, you can select up unused quantities at material centers, outside earnings, thrift shops, and severa distinct assets. All you need is to apply a chunk of fabric that has been washed and dried.

You can reduce6x6 pieces of fabric. Make sure it's a breathable one.

Step 1: Sew Pillowcases

Place the 2 squares of cloth with the proper sides together. Sew three edges collectively. You can use approximately a three/8" seam allowance, however don't strain too much about it. A little tons less received't harm.

Sew the fourth aspect collectively, however don't give up it; leave about 2" unsewn and open. You can do this stitching with a device or via the use of hand.

Using that setting out, flip the pillowcase right-side out. Use a chopstick, knitting needle, or crochet hook to push the corners out.

Step 2: Fill Pillow with Herbs

Loosely fill the pillow with a few fiberfill to provide it a few form. Don't overstuff it; it should be flat enough to suit in a regular pillow. Then, upload more than one tablespoons in conjunction with your chosen herb combination.

Step 3: Finish it up

Tuck the raw edges of your organising in the pillow and stitch that beginning closed. Clip off the strings and make sure to do away with any herbs that might be sticking to the outdoor.

Your pillow is now organized for use or for gifting. Make certain you commonly tell the recipient which herbs are interior and why you picked those.

•Lotions for personal care

The non-public care enterprise is roaring louder than ever. It's no wonder why it has made millionaires out of such quite a few small corporations and marketers who dared to try their hand at it.

Visiting a community drug maintain offers you masses of facts on how much room there is inside the market if you need to take. Most of the lotions you discover at drug stores are almost 70% water. Take a test the normal expenses, and honestly expect that you can be making a much more herbal, powerful lotion for 20% of the price.

You ought to make a lotion that gets rid of those bizarre chemical compounds listed within the components and get preserve of a markup of 500% to 800%.

Contrary to what you will probable anticipate, the device used to make lotion on a small scale isn't always high-priced. When you're sincerely beginning out, you could need to use a number of the kitchen gear you have got already got, which include a thermometer and stick blender, to maintain fees down. As your employer grows, you'll be capable of upload greater gadgets that might assist you improve general overall performance.

You may additionally need to attempt developing a skin-soothing lavender lotion, for starters. The materials are smooth to go back by way of way of the use of, and the method is even less complex. Here's a way to do it.

•three/four cup oil, consisting of candy almond oil

•2 tablespoons shaved beeswax (keep away from the use of waxes made with petroleum merchandise)

•30 drops lavender crucial oil

•1 cup water

•One sixteen-ounce glass jar or two 8-ounce glass jars for storing lotion

Step 1: Heat

Add the oil and beeswax to a small saucepan and warmth over low warm temperature till wax dissolves. Remove from heat right now.

Allow it to relax for 1 to two mins. Then, add the lavender essential oil. Once it's mixed well, set it aside.

Step 2: Blend

Pour water into a blender and start mixing on excessive speed with the lid on and the hollow in it left open. With the blender walking, slowly upload the oil-wax mixture. It will start to emulsify as you hold pouring in oil.

Step 3: Store

Pour the lotion into glass jars. It will final approximately 6 months and is remarkable stored at cool temperatures, or even in the fridge. This recipe makes about 2 cups.

There's additionally a shortcut you can try if you revel in you are too busy strolling topics to make the lotion from scratch. The key's using bulk skin care bases, which can be pre-mixed. You can discover hand and frame creams, shower gels, conditioners, shampoos, soaps, foot creams, doggy shampoos, lip balms, massage oils, body butters, and additional.

Be certain to test the materials of these bases earlier than buying them. If you need to lean inside the course of an all-natural kind of product, you could enjoy extra cushty doing it all from scratch. But if the bases are natural as properly, you then without a doubt need to be proper to move and try them out.

They can be to be had unscented, so that you can certainly upload your own lavender oil to the premixed base. Then, all you need to do is bundle your product, label it, and you are done for the day. You can use this to do A/B sorting out of your product, as properly. This approach attempting first-rate variables to look which suits first-rate.

You can strive shopping for a few gallons of the bottom, preparing the combination, and taking it to a local Saturday marketplace. You also can publish it on-line in locations like Etsy.

Chapter 14: Packaging

Whether you're selling creams, shampoos, shaving creams, or lip balms, there's one thing you can't skip, and it's far the packaging.

In a aggressive market, packaging is a lot greater than truly putting your product in a container — it's an possibility to "wow" your patron. And to preserve them coming. Make beginning your package deal an revel in. Showing your clients which you went the extra mile will make a tremendous impact. This can be done in small but wonderful gestures together with a thank you card or growing eye catching outside packaging layout.

So, why do those information rely range loads? They help you stand glad with the crowd.

If you think about it, how usually have to procure some component you don't need truly due to the reality the product's packaging is attractive or suits your character? If you haven't, you then genuinely definately ought to have visible ultimately the magnetic enchantment that packaging has on a pal of yours.

This is the reason why big brands like Apple or Glossier supply loads hobby to the information of their programs.

Your product's packaging is often the primary effect and first difficulty of contact collectively together along with your capacity customers. Combining suitable packaging with a beneficial product is the proper technique for a faithful consumer of your emblem.

For creams and lotions, maximum bottles are to be had in 2, four, 8, and 16-ounce sizes, however beyond that, you want to pay specific interest to how they may be provided and taken to the client. You might sense tempted to get a smooth plastic bottle, stick a label to it and call it an afternoon, but I inspire you to head in addition.

Think of staggering discipline format, putting together mini collections with one of a type products, inclusive of gambling cards to the consumer, or having a stunning label that catches the attention.

If you observed you've got a few mild layout skills, you could attempt designing your labels and brand image in Canva.Com, a unfastened software application software with hundreds of templates you may use. It is fantastically clean for beginners, and it has plenty of options to gather concept.

If your format capabilities are not too pinnacle, I must propose hiring a fashion designer that will help you out. Online and in social media, you could discover an great quantity of designers providing label design, package deal deal deal format, branding visuals, social media visuals, and plenty extra. You can use freelance systems, like

Upwork, to discover upcoming freelancers that could use the assist.

Apart from the format, you'll need to embody all things criminal. It need to have a internet web web page address or distinct contact records to make it easy for purchasers to contact you to re-order. Also, include any data required via the FDA, inclusive of a listing of substances, listed so as from the maximum to the least quantity.

Most creams have approximately 70% water, so distilled water could be the primary trouble indexed on the label.

The FDA additionally requires which you listing the burden in oz.. And your name and cope with or a employer call and address. To check current FDA policies and hints protecting this topic, go to them at: www.Fda.Gov/cosmetics

•Acne moisturizer with lavender

Plenty of humans, young adults and adults protected, be tormented by manner of the uncomfortable presence of pimples in extremely good factors of their our bodies. The suitable records is that you could assist them. Because lavender is antibacterial and antiseptic, it is often used to assist those tormented by pimples and different pores and pores and skin problems together with eczema.

One manner in which lavender oil may have an effect at the pores and pores and skin is via the usage of changing sebum manufacturing. When the pores and skin is dry, the oil glands at the pores and pores and skin begin overproducing sebum to compensate for the lack of moisture. This is terrible because of the reality the excess sebum clogs the pores and ends within the formation of zits. With ordinary use of lavender oil, the pores and pores and skin will live moisturized, gentle, and colourful -

the sebaceous glands obtained't be brought on.

Even in case you don't have dry pores and skin, pressure can also make the oil glands circulate into overdrive. This is exactly how stress can purpose pimples flare-ups. When the body is alarmed, in any other case you feel aggravating, you sweat extra. The pores and pores and skin produces hundreds more oil. Lavender's strain-relieving homes are a powerful relaxing

It can also penetrate into the pores, killing micro organism and getting rid of dirt. Additionally, it blessings the pores and skin as it has effective antimicrobial and antioxidant trends.

Among extraordinary benefits, lavender oil is famous for its antifungal houses. When someone suffers from eczema, their pores and pores and skin will become dehydrated, itchy, and scaly. The vital oil can revoke the ones signs and symptoms and symptoms, lessen infection, and keep the situation on pinnacle of factors.

Combine it with the right provider oil, in this situation, jojoba oil, and you've had been given a herbal product that has no harsh chemical materials to worry about. The jojoba oil is just like the natural sebum discovered in the pores and skin.

•eight oz. Jojoba oil.

•0.08 oz.. Lavender crucial oil.

•0.04 ounces. Tea tree essential oil.

•Lavender bathtub oils

Baths are supposed to be fun, soothing, and cleansing. Adding lavender oils can maximize rest and may assist humans sleep better. It can dial up your tub, adding strain-relieving, thoughts-soothing, muscle-enjoyable strength.

There also are severa research studies that show lavender facilitates loosen up kids with ADHD (Attention deficit/hyperactivity disorder) in case you want to get an top notch night time time's sleep.

You is probably questioning that lavender bathtub oils refer essentially to including the lavender oil proper away into the water after the bath is entire. Well… no longer exactly. When it comes to placing vital oils within the bathtub, recall — oil and water don't combination, this is to mention the vital oils are not water soluble.

When you climb right proper into a bath under those situations, the small drops of oil can adhere on your pores and pores and skin (and distinctive easy areas) really as in case you'd completed the undiluted oil without delay. This must worsen or burn your pores and pores and skin.

This is why you want to constantly integrate critical oils with a company oil first.

You don't need the lavender oil just to take a seat down on top of the water; you want it dispersed inside the direction of. The super manner to do this is to combine critical oils with a agency oil first, like coconut, olive, sunflower, or jojoba.

For a single tub, three to 12 drops of critical oil in a tablespoon (15 ml) of carrier oil is enough to create an fragrant, healing tub.

Here are a few wonderful recipes you could attempt depending on what you're centered on with the lavender bathtub oil:

Muscle-Soothing Bath

Tired or overworked muscular tissues? This trio can dial up flow into and dial down pain.

•five drops marjoram (additionally called candy marjoram)

•4 drops lemongrass

•three drops lavender

Relaxation Bath

This calming and exciting blend can reduce pressure and get you organized for bed.

• 5 drops lavender

• 4 drops chamomile (German or Roman)

• three drops frankincense

• Lavender tub salts

As with lavender bathtub oils, occasionally you need to take the time for your self and revel in a chilled tub. Another alternative that may assist with this is lavender bathtub salts. I saw a nice gift % of bathtub salts in a store the alternative day – the shape of keep that sells a piece bit of this and a chunk little little little bit of that: trinkets and gadgets and homewares. The fee tag?

Twenty-eight dollars!

Can you calculate the markup for that?

Lavender bathtub salts are top notch easy and cheap to make your self, and you can get all of the additives you want at your network grocery keep.

Here's what you can need:

•Epsom salts.

Find Epsom salts inside the toiletries aisle of the grocery keep, normally in the 'medicinal' segment. Look on the lowest shelf and the top; they may be hidden sometimes. Epsom salts are surely magnesium sulfate. When delivered to water, the 'salts' dissolve into magnesium and sulfate ions. It is the magnesium specially that is supposed to have relaxing houses as it is absorbed into the pores and skin.

•Sea salt

You probably have already got this aspect on your pantry. Just make certain it's no longer easy vintage table salt. For making bath salts, actual sea salt or kosher salt is high-quality so that you but get all the minerals placed in unprocessed salt.

•Bicarbonate of soda

Bicarbonate of Soda is traditionally applied in a tub for its detoxifying houses. Because its PH is slightly alkaline, bicarb can neutralize acids, that is why it's used as a bath salt element.

•Lavender crucial oil and Dried lavender (optionally available)

You can also encompass a drop orof blue and pink meals coloring. It offers the salts a purplish tone that appears extraordinary whilst packaged.

Here's how you could make your lavender salt at home.

Grab a medium size mixing bowl and pour 1 cup of Epson salt inside the backside of the bowl. Pour a tablespoon of baking soda proper on top of the Epson salt. Mix them collectively.

If you pick out to function in meals coloring, now may be the time to do it. You can add one drop of blue and one drop of pink and blend nicely to get a moderate red lavender shade.

Add the lavender oil some drops at a time, blending thoroughly. If you decided to apply the dried lavender, add it to the aggregate and stir.

Once it's organized, you could keep it in an air-tight jar or vicinity.

Remember to take into account the package format while labeling and storing the lavender bathtub salts! People love getting these as a gift for others, so that you can keep that during mind at the same time as promoting them in your customers.

•Lavender frame butter

Lavender body butter seems delicious — exactly like frosting. But I wouldn't endorse eating it, even though it has the phrase "butter" in its call.

Many industrial body butters use harsh chemical substances as thickeners that may be risky on your skin or can dry it out, it is the exact opposite of what you may need. This is why a natural, artisanal lavender body butter can be an outstanding extraordinary-supplier for a lavender grower.

You can make it through combining shea butter, coconut oil, jojoba oil, and lavender important oils. Shea butter comes from the seeds of the fruit of the shea or karite tree, and it is high in vitamins A and E. No surprise this butter will make any sort of pores and skin revel in smooth and silky.

Here's how you can do it:

Melt the coconut oil and shea butter collectively, then combo inside the jojoba oil and important oils. Once it's prepared, region it in the fridge to harden (both in a mixer bowl or a bowl you may use hand beaters in) for 1-2 hours.

Remove from the refrigerator and whip for approximately 10 mins, or till the frame butter looks like frosting.

You also can begin from an unscented base, which you may discover in places like www.Bulkapothecary.Com.

Try superb options and observe what works best. Remember to percent it up nicely and upload all of the vital records to the label.

•Lavender lip balm

Like the rest of the private care merchandise, lavender lip balm is an superb desire to must your catalog. It's smooth and reasonably-priced to make, and you can sell it on line, to community grocery stores, or on the Saturday market.

It is the right item to have earlier than the cold months start, so you want to time it with the season for a larger impact for your clients.

Here are a couple of recipes that you may observe to make your personal lip balm. Using super additives will offer your customers extra options to pick out out from.

Lavender Honey Homemade Lip Balm

•2 Tbsp of coconut oil.

•1 Tbsp of shea butter.

•1/2 tsp of uncooked honey.

•1 Tbsp of sweet almond oil.

•2 Tbsp of beeswax

•15 drops Lavender critical oil

•five drops Frankincense crucial oil.

To store it, you may use lip balm tubes or small recipients.

If you pick out to make it with the tubes, the first step is to put off the lids from the lip balm tubes and solid them upright with a large rubber band.

Then, gently soften the coconut oil, shea butter, honey, and beeswax in a double boiler (or in a glass bowl set on top of a sauce pot of simmering water). Remove from the heat and stir inside the sweet almond oil and essential oils.

Quickly pour the melted oil into the upright tubes. If you'd favor to bundle it in super recipients, have a look at the identical way and pour the aggregate because of this. You need to pass very quickly earlier than the oils start to set.

Allow the lip balm to set and cap the boxes. The heady scent can be intoxicating.

Here's each other recipe that you could try. This one makes use of peppermint rather than honey.

•1/4 cup of beeswax beads (or candelilla wax for a vegan possibility)

•1/4 cup of shea butter

•1/4 cup of coconut oil

•10 drops of lavender essential oil

•10 drops of peppermint critical oil

The gadget is basically the equal.

This recipe will make sufficient to fill about 15 x 15ml glass pots. You should fill one 15ml glass pot and keep the remainder in a larger glass jar. When you are organized, top off your lip balm pot, soften it down, top off and leave to set.

Label your little packing containers and positioned them up for sale. It gained't be long till they are lengthy beyond.

•Lavender linen water

In the wintry climate months, some of us are caught indoors heated homes most of the time. Don't get me wrong, I'm grateful for heating even as the weather is cold, but I do lengthy for a heat, summer time breeze to go with the flow past my curtains.

The summertime air smells like flora, grass, and sunshine. So in late summer season, as soon as I harvest my lavender and set my linens out to dry on the clothesline, I circulate slowly into sparkling, flowery sheets at night and burst off to dreamland. On the lengthy winter nights, my linens are well, blah. No sunshine to warmth them and no summer time air to guide them to scent like heaven.

My little taste of summer time comes in a bottle of lavender linen water that I can spritz on my sheets even as making the mattress. Or use in my iron instead of easy (stupid!) water. Or to clean up towels, pillows, dog beds – you call it!

And if you upload this for your product catalog, your customers will sense the identical way. This is every other tremendous product to characteristic right earlier than wintry weather. And who is aware of, perhaps it's going to stay the complete one year. It's easy to make, and the outcomes are thoughts-blowing.

To make it, add 4 cups of distilled water, 1/4 cup witch hazel, and 25 drops of lavender important oil to a bottle and shake. It's a smooth as that to deliver a touch summer time again into your days. And deliver you off to dream of summer season nights.

Some recipes don't consist of the witch hazel. Instead, they use a cup of unflavored vodka (acts as a preservative and emulsifier to help the oil and water mix together). It also can be made with dried lavender buds rather than vital oil. Just upload one cup of dried lavender buds to two cups distilled water in a saucepan, simmer for 10-15 minutes, after which permit it to steep till cooled down.

Strain the cooled liquid via a cheesecloth or exquisite mesh strainer and upload ¼ cup of alcohol as a preservative.

My advice is to attempt splendid recipes and see what works excellent on your case. When deciding on your packaging, don't forget that it's far typically bought in four, 8, or 12-ounce spray bottles.

•Lavender shampoo

All-natural hair products price big dollars in shops. But growing a DIY herbal shampoo is highly brief and easy. The way takes 10 mins max, and in case you are concentrated on customers who need to preserve a herbal life-style, freed from chemical substances, they may love your lavender shampoo.

Dr. Bronners castile cleaning cleaning soap is crafted from a vegetable-base, making it vegan and herbal. Unlike traditional shampoo supplied in stores, castile cleaning cleaning soap is an inexperienced difficulty.

Homemade shampoo is free of the fillers and chemical irritants in business shampoo.

•half cup distilled water

•1/2 of cup of liquid castile soap (unscented)

•1 teaspoon of jojoba oil (grapeseed, apricot, sweet almond, avocado, or each extraordinary mild vegetable-primarily based without a doubt employer oil works too)

•Lavender crucial oil

•Amber glass bottle dispenser

Combine the elements in a reusable amber glass bottle dispenser with a pump. To fill a 16-ouncesbottle, double the recipe measurements above. Add the castile cleansing soap and oils in first, determined via the water, for fewer bubbles. Swirl the contents of the bottle to mix.

Lavender products for the kitchen

We understand lavender for its unforgettable heady scent and colorful shade, however have you ever ever tasted it? Even handiest a tiny chew? Culinary lavender is specific and bold, making it highly flexible in recipes and super for stepped forward everyday cooking.

Here are a few records you likely didn't recognize:

•Lavender stems are flavorful and may be used as a drink's stir stick or a skewer for kabobs.

•Lavender flora and grape vines are an superb pair. When planted together, the lavender lets in ward off insects and viruses, making them a great duo for natural farming.

•Edible lavender continues to be pretty new to america cooking scene, or even into the 1990's lavender wasn't implemented in everyday recipes.

You might be wondering, is all lavender suitable for eating?

Technically sure, but only a few lavender types are considered actual culinary lavender and characteristic a delicious flavor. Royal velvet and Provence kinds are labeled to be culinary lavender and are satisfactory for cooking. Other sorts, like Grosso, are extra generally used as decoration, bouquets, and for vital oils, not inside the kitchen.

Whether it's sparkling or dried, lavender must be handled much like a few other herb. If stored out of direct slight and in an hermetic and dry container, lavender will live flavorful and aromatic for 1-3 years. Using lavender beaten, extra or less chopped, or whole in a dish can upload excessive taste and can be applied in quite a few techniques, like in drinks, meats, or desserts.

Even although it's scrumptious, dried lavender is difficult to return by means of using the use of on the grocery shop. It's the right hollow within the marketplace for small growers looking to expand their lavender industrial business enterprise.

The 5 exquisite types for culinary use are:

•Lavendula augustifolia "Buena Vista."

•Lavendula augustifolia "Folgate."

•Lavendula augustifolia "Hidcote Pink."

•Lavendula augustifolia "Melissa."

•Lavendula augustifolia "Royal Velvet."

Here you'll find out a few high-quality lavender culinary products that you may with out problems make and promote. They can all be used within the kitchen, and they're capable of deliver dishes a flavorful contact.

•Herbes de Provence

A famous herbal mixture, Herbes de Provence is considerably used as an all-motive seasoning mixture for meat and vegetable dishes. It is normally a aggregate of lavender, thyme, savory, oregano, rosemary, and marjoram.

It also can have chervil, oregano, mint, parsley, and tarragon.

If you've ever wanted to tour to the South of France however haven't been capable of swing an plane charge tag, Herbes de Provence would probable belong for your grocery listing. A staple in French and Mediterranean cooking, this floral and slightly woodsy herb combination can shipping your taste buds to a French café with a shake of a spice jar.

To make it, aggregate collectively a mixture of dried herbs. You can strive it out to your dishes and come up with an authentic recipe to help you stand out to your customers' eyes.

You can also actually sell the dried lavender buds and encourage your clients to make their own mixture!

•Lavender tea

This has in all likelihood been roaming the decrease once more of your thoughts due to the reality that I delivered up the culinary trouble of lavender. Lavender tea has been brewed for hundreds of years to assuage and loosen up the involved system.

The excellent advantages of lavender tea can also moreover include exciting the frame, reducing muscle spasms, promoting healthy digestion, and assisting sleep. It might also assist dispose of inflammation, stability temper, heals the pores and skin, and soothes persistent ache.

Lavender tea comes from the lavender buds, the small pink bundles. When you drink lavender tea, it could offer comfort from insomnia, excessive anxiety, gastrointestinal disappointed, pores and skin irritation, and headaches. Regardless of what ails you, the advantages of lavender tea will possibly be able to help.

Making lavender tea at home is a clean way, related to great glowing lavender buds and water, in spite of the truth that some people enjoy mixing the tea with honey, chamomile, or possibly different varieties of tea along side lavender milk tea.

Instead of absolutely promoting your lavender buds in bulk, why now not process them to function charge (about four hundred% extra!) with the aid of selling lavender tea bags.